THE SINAI

THE SINAI
A Physical Geography

Ned H. Greenwood

University of Texas Press, Austin

Copyright © 1997 by the University of Texas Press
All rights reserved
Printed in the United States of America
First edition, 1997

Requests for permission to reproduce material from this work
should be sent to Permissions, University of Texas Press, P.O.
Box 7819, Austin, TX 78713-7819.

♾ The paper used in this publication meets the minimum re-
quirements of American National Standard for Information
Sciences—Permanence of Paper for Printed Library Materials,
ANSI Z39.48-1984.

Library of Congress Cataloging-in-Publication Data

Greenwood, Ned H., 1932–
    The Sinai : a physical geography / by Ned H.
Greenwood. — 1st ed.
        p.   cm.
    Includes bibliographical references and index.
    ISBN 0-292-72798-4 (alk. paper). — ISBN 0-292-72799-2
(pbk. : alk. paper)
    1. Physical geography—Egypt—Sinai.   I. Title.
GB332.G74   1997
915.3'1—dc20   96-27618

Contents

List of Illustrations		vi
Preface		vii
Place Name Types and Derivations for Sinai		xi
1.	Sinai: Search for a Geographic Entity	1
2.	Plate Tectonics and the Geology of Sinai	11
3.	Geomorphology and Drainage	26
4.	Weather and Climate	51
5.	Soils of Sinai	70
6.	Biogeography of Sinai	88
7.	May They Eat Lamb in Paradise	114
	Bibliography	133
	Index	137

List of Illustrations

Preface Maps
Physical Sinai   ix
Cultural Sinai   x

Figures
1-1. Crossroads banzine station   3
1-2. Governorates   6
1-3. Important settlements   7
1-4. Maaza Bedouin   9
2-1. Cretaceous sea-land arrangements   16
2-2. Domal uplift and rifting   17
2-3. Divergent plate boundary   18
2-4. Surface geology   21
2-5. Nubian sandstone   23
3-1. Geomorphic regions   27
3-2. Cross-section of geomorphology   28
3-3. Cuesta at Gebel Halal   30
3-4. Satellite imagery, Tih Plateau   32
3-5. Sand valleys   34
3-6. Dividing Valleys   35
3-7. Tectonic cross-section   36
3-8. Aqaba Foreshore   38
3-9. Intrusive gabbro dikes   39
3-10. Gebel Serbal   41
3-11. Feiran-Serbal area   42
3-12. Gebel Musa area   44
3-13. Summit basin   45
3-14. Sinai drainage   47
4-1. July mean temperature   54
4-2. January mean temperature   55
4-3. Pressure and wind, January   56
4-4. Pressure and wind, July   57
4-5. Annual precipitation   58
4-6. Aridity   61
4-7. Sinai's climates   64
4-8. Climatic graphs   65
4-9. Water budget diagrams   67
5-1. Soil regions   72
5-2. Cultivable soils   74
5-3. Bazooka rockets   75
5-4. Students examine soils   76
5-5. Entisol profiles   77
5-6. Aridisol profiles   84
6-1. *Acacia raddiana*   91
6-2. Vegetative groups   97
6-3. Common jerboa   105
6-4. Nubian ibex   108
6-5. Western reef heron   110
6-6. Desert monitor   111
7-1. Africa looks at Asia   116
7-2. Salaam Project   121
7-3. Arab fishermen   126
7-4. Basic house   128
7-5. Wind surfing   132

Tables
2-1. Geologic column   14
3-1. Sinai's principal peaks   40
4-1. Mean daily radiation   52
4-2. Receipt of sunlight   53
4-3. Climatic stations   62
4-4. Universal thermal scale   63
5-1. Entisol laboratory analysis   78
5-2. Aridisol laboratory analysis   85
6-1. Higher plants   92
7-1. Fisheries data   125

Preface

Chapter 95 of the *Qur'an* (*Koran*) begins, "By the Figtree and the Olive, Mount Sinai and this Safe countryside, We have created man with the finest stature; . . . " My introduction to the "figtree and olive" came in 1989. In January of that year I went to Ismailia for the first time, eager to begin field work and research in the Sinai. In my naivete I was quite unaware that the transfer of my Senior Fulbright Research Fellowship from the University of Alexandria to Suez Canal University had been manipulated by Dr. Ahmed Ismail Khodeir, President of the latter institution. President Khodeir, a member of the Fulbright Selection Committee in Egypt, had viewed my training in physical geography, soils, and water resources and decided to put me to work helping set up the laboratory and field research programs for the new Faculty of Environmental Agricultural Resources at the El Arish Campus of Suez Canal University. This was a great bit of serendipity which facilitated a wide exposure to the geography of Sinai. I am indebted to Dr. Khodeir for this opportunity and the personal encouragement which he offered along the way.

For the reader less than familiar with Sinai there are apt to be problems with specific areas of regional geography. Place names are not always easy to integrate with a ready accessibility. Wherever possible there is a cross-reference between given place names and the nearest map or illustration found in the text. However, there will always be those names that slip through the grid. To mitigate this loss two maps are included with this preface for quick reference. The first map (p. ix) deals with the physical locations such as wadi (dry stream bed), gebel (mountain), naqb (pass), and ras (headland). Terms such as ain (spring), bir (well), seil (torrent bed), moyat (pond), etc., are more localized and apt to appear on specific illustrations. The second map (p. x) shows general features such as highways, highway passes, and major towns.

Finally, I would like to express my appreciation for those who have been so helpful and patient in the preparation of this book—to the reviewers, Dr. Joseph Hobbs and Dr. Robert Gabler, who gave much time and encouragement at critical points during the revisions of the manuscript; to Laura Edgeman, who gave freely of her time and skill to the preparation of the map illustrations; and to my wife Alverta, who did much of the power keyboard work. Thanks to Shannon Davies, the Sponsoring Editor at the University of Texas Press, who initiated the project and relentlessly carried through the long months of preparation. Thanks to my colleagues at San Diego State University, whom I called upon for help with numerous details that invariably cropped up in the manuscript preparation. Thanks to my many friends and mentors at Suez Canal University and in Sinai from whom I learned so much. As the reader progresses through the book the contributions and encouragement of this latter group will be quite evident.

As this project began to evolve I was considering a comprehensive geography of Sinai, but the magnitude and intricacies of the cultural aspects were so vast that the impracticality if not the impossibility of such an undertaking became increasingly apparent. With the encouragement of my Editor and the Faculty Advisory Committee at the University of Texas Press, I decided to focus on the physical aspects of that geography and in finale relate them to the immediate problems of mankind in the present environment and the potential for sustainability. Hence the wish (and title of the final chapter) borrowed from an ancient Bedouin greeting/benediction—"May we all eat lamb in paradise."

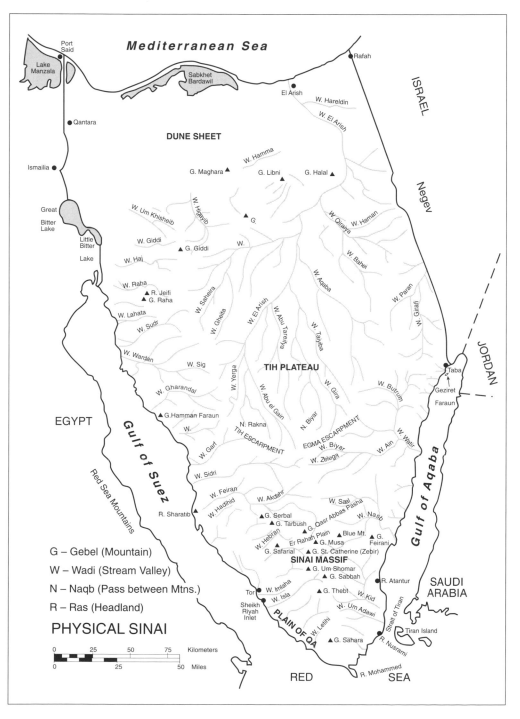

Physical Sinai

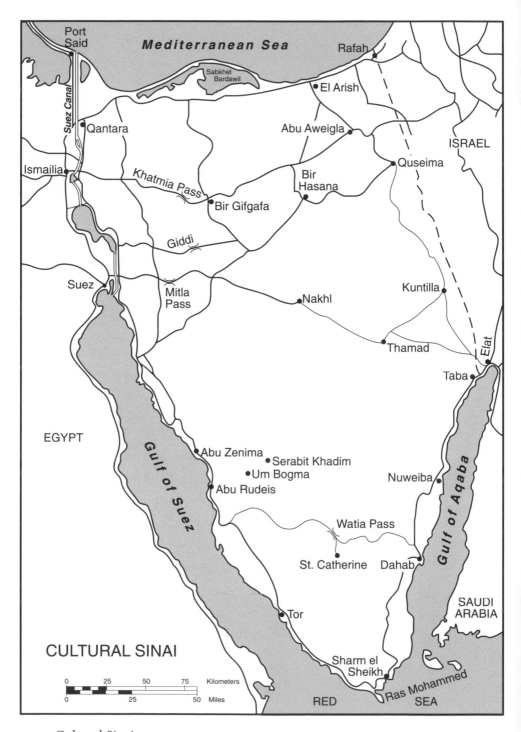

Cultural Sinai

Place Name Types and Derivations for Sinai

Generic Place Names

Ain   natural spring of water, Ain Akhdar=Green Spring.

Arish   shelter made from palm fronds, Wadi el Arish.

Bilad at Tor   Land of Mountains, original name for south Sinai.

Bir   well.

Bueb   door, like a narrow place in a wadi.

Dhayga   narrows, as where a wadi is constricted by mountains.

Galt   cistern.

Gebel   mountain, also commonly shown as jebel, Gebel Hadid=Iron Mountain.

Jawra   flat plain surrounded by sand dunes.

Maghara   cave, Gebel Maghara=Cave Mountain.

Minit at Tor   Port of the Mountains, now simply rendered "Tor."

Moyat   water, as in a pond.

Naqb   pass between mountains, Naqb el Mirad=Pass of the Watering Place, Naqb Buderah=Pass of the Sword, Naqb Hawy=Pass of the Wind.

Qa   plain.

Ras   peak or headland, Ras Ahmar=Red Peak, Ras Mohammed= Headland of Mohammed, the tip of the Peninsula extending into the Red Sea.

Sabkha   salt flat.

Seil   the lower reaches of a wadi or dry torrent bed, often where several wadis come together.

Sharm   break in the coastline, a bay, Sharm el Sheikh=Bay of the Sheikh.

Shust   summit.

Sidd   dry waterfall.

Tawil   tall, pointed peak.

Tell   hill, often a ruin.

Themila   area of shallow groundwater.

Wadi   stream course, usually dry, Wadi Mukateb=Written Valley, inscriptions.
Yirga   twisting wadi, Ain Yirga=Spring of the Twisting Wadi.

Zoomorphic Place Names

Ghazala   gazelle, Wadi Ghazala=Gazelle Valley.
Halal   goat, Gebel Halal=Mountain of the Goats.
Humr   fruit of the *Nitraria retusa,* a nitrogen-fixing plant with thorns and fleshy leaves.
Ifai   viper, Wadi Um Ifai=Valley of the Mother of Vipers.
Kuntilla   a grass, *Bromus sinaicus,* Sinai brome.
Sa'al   acacia tree, Wadi Sa'al.
Sidri   thorny shrub, *Zizyphus spina-christi.*
Sulaf   herb, *Capparis cartilaginea.*

Historic and Legendary Place Names

Badiet at Tih   Desert of the Wandering (Children of Israel), most of central Sinai.
Dahab   gold. Thought to be the port for Solomon's gold coming from Ophir, Dahab is an unlikely spot for a port serving Jerusalem.
Deir el Arbain   Convent of the Forty (forty martyrs killed by Bedouin raiders).
Magad Sayidna Musa   Sitting Place of the Lord Moses (a red granite rock near Watia Pass).
Sabkhet el Bardawil   Lake Baldwin (named for Baldwin I, King of Jerusalem, who died here in A.D. 1118).
Serbal   Baal's Vineyard, as in Gebel Serbal.
Sikket Shoeib   Path of Jethro, near Gebel Musa and St. Catherine.

Miscellaneous Place Names

Abu Tarfa   Father of Tamarisks (a large grove of manna-bearing trees).
Um Shomer   Mother of Darkness (a peak in the Sinai Massif composed of dark igneous rock).
Um Dud   Mother of Worms.

1. Sinai: Search for a Geographic Entity

Sinai has for centuries been a land that has stimulated scholarly endeavor and fostered a widespread lay interest. However, it is the recent decades of concentrated political and military conflict that have eclipsed the age-old need for better knowledge. Yet for all its importance there are few if any convenient sources to which the modern reader in the West can turn for a concise and comprehensive look at this land as a total entity. This work attempts to provide a new focus on Sinai's physical geography.

With paved roads and motor vehicles it is possible to travel from Ras Mohammed, at the southern tip of Sinai, to Qantara, in the northwest on the Suez Canal, then from west to east all the way to Rafah in one very long day. In a week one could travel every road in the peninsula, yet it would take decades to achieve a real familiarity. For a Westerner it may be impossible to ever fully know Sinai, but there is a powerful fascination with its natural environments and peoples—in a word, its geography. As observed by American friends after we had made a camel trek in a snowstorm to the Blue Mountain, where a mad French artist had painted vast areas of brown desert rocks an improbable blue: "Sinai is a land of fascination and frustration." While it is a fascinating land it has also been a greatly disputed land. Over the past half-century it is doubtful that any real estate on earth has been more negotiated and fought over than this small triangle of desert that links the great continental masses of Africa and Eurasia like a short umbilical cord. This juncture of continents is as pivotal politically and culturally as it is tectonically. The inhabitants brag that it is the "crossroads of continents and meeting place of seas." With a bit of juggling one could easily substitute "meeting place of humanity." Through millennia diverse floods of humanity have washed over its arid wastes, pushed and pulled between Pharaonic Egypt and Assyria-Babylon. Between Arab Egypt and Hebrew Israel. Between Christian, Jew, and Muslim. Nor should it be forgotten that Greeks and Romans struggled for imperial control, as did the British and French. Even the United States and the Soviet Union have been tugged upon by Sinai's lodestone. Through the course of history the turmoil of Sinai has been more often than not tied to that of Egypt, whatever its ruling power.

In a Name

In biblical times Paran, not Sinai, was the name for the peninsula. Nineteenth-century Arab Egyptians were apt to call it the Way to Hejaz. This is a reference to the Hejaz province of Arabia, which contains the holy cities of Islam: Mecca and Medina. Such a reference is somewhat disparaging, indicating that it was little more than empty space one had to cross to reach the important goals, the quest of the haj or pilgrimage. Another commonly used term, "Shibh Jazirat Sinai," was more definitive and carried no backwater overtones. It simply means "Peninsula of Sinai."

The word "Sinai" is so old that its derivation is uncertain. Some believe it was derived from "sen," meaning tooth, which describes the shape of many mountains found there. Others claim that it comes from "Sin," the Moon God of Sumeria, who was also the Lord of Wisdom and Father of Gods. The worship of Sin was favored by the Nabataeans in Sinai. Whatever its source, the name Sinai is now unequivocal. It means only one place and its boundaries are relatively fixed.

Even as the name is unequivocal, the physical regionality is concise and easily defined. Except for the northeastern boundary, Sinai would be as easily demarcated as an island. But along that sector the artificiality of man-made boundaries is glaring. The close physical and cultural relationships to Israel's Negev were not erased by the drawing of an artificial line across the sands. Even decades of more-or-less continuous war have not destroyed these cultural and geographic ties. The Bedouin of the Negev still look to their cultural roots in Sinai.

Physical Regionality

The details of physical regionality are mapped in Chapter 3 (Figure 3-1), but this chapter needs to consider a few salient divisions. From north to south the eight main regions are: (1) the Dune Sheet of the Mediterranean Littoral; (2) the Insular Massifs, containing gebels (mountains) Maghara, Yelleq, and Halal; (3) the sandy deserts of the Suez Foreshore; (4) the gravelly and rocky Tih Plateau, which includes the main drainage of the Wadi el Arish; (5) the Dividing Valleys, which lie between the Tih Plateau and the Sinai Massif;

Figure 1-1. Crossroads banzine station near Gebel Libni, looking south across the flat, nearly featureless sand plain toward Bir Hasana. In the far distance can be seen the foothills of Gebel Yelleq.

(6) the Sinai Massif; (7) the Plain of Qa; and (8) the Aqaba Foreshore. These regions are each readily defined in terms of abrupt breaks in the landscape.

It is quite possible that Sinai is as unique tectonically as it is theologically. The events of crustal plate movement, the push and pull of the large African and Eurasian masses combined with lesser adjustments within and between the Levantine and Arabian plate fragments have created an intricate pattern of lands and seas. In the north the land is washed by waters of the temperate Mediterranean. In the south the peninsula is grasped by the arms of the tropical Red Sea. Neither sea does much to alleviate the aridity imparted by the Atlantic Subtropical High Pressure System and enforced by the great land space of Saharan Africa. The Mediterranean fringe of Sinai suffers from the environmental pollution and declining ecosystem complexity that plague the entire basin. Likewise ocean shipping and petroleum extraction in the Suez arm of the Red Sea are problematic, yet there remains a strong contrast between the rich life in the sea margins and the intense deserts of the adjacent coastal plains.

On the land itself the contrast is between flat, nearly featureless plains and rugged, barren mountains. In the north, mountain remnants like Gebel Maghara rise precipitously from alluvial plains.

In the south, the Sinai Massif dominates surrounding plains with range after range of exposed crystalline mountains.

On the whole there is less contrast between lifeforms than geologic forms. Nowhere is the vegetative cover really luxuriant. Even the oases tend to be mere strips of life seemingly meant to emphasize the barrenness of surrounding deserts. Subdesert grasses with scattered scrub communities are widespread. These appear lifeless through much of the year, responding briefly to late winter rains and runoff. Then the shrubs take on a lighter color, but most do not show a vivid green, rather a lighter hue of olive drab. To the casual observer there seems to be little contrast, but drab shrubs and annual grasses hold a surprising diversity.

Land wildlife provides less contrast than vegetation, largely because it is seldom seen. Reptiles are not as common as in comparable desert areas of North America. Larger mammals like the ibex and jackal are almost never seen. Mostly it is the domestic animals, the "giants" (camels) and the "small cattle" (sheep and goats), that lend an animal presence to the landscape. Bird life, fortunately, is found in considerable variety, especially during the migratory seasons of spring and fall. In spite of an overall paucity, lifeforms are so important to the culture that they are prominent in folk stories and frequently show up in place names.

Place Names

In 1869 British explorer and Orientalist Edward Henry Palmer, under the auspices of the Palestine Exploration Fund, traveled extensively in Sinai to collect correct place names and develop an accurate geographic nomenclature. This was the first systematic attempt to render Sinai names in English. Palmer, like the rest of us, was confronted with the problem of transliteration—how to render the Arabic phonetic sound in the English alphabet. Unfortunately even now there is no standard system. The rendition of names varies greatly. For example, the town at the center of the Tih Plateau is variously given as "Nakhl," "Nekhel," and "Nukhl."

The generic name for mountain is given as "gebel," "jebel," or "jabal," with the Bedouin pronunciation falling somewhere between the English "g" and "j." Pilgrim is rendered as "haj," "hajj," or "hogg." There is no definitive system, only workable systematics. In the following pages there will be an attempt to render terms consistently and as simply as possible. "Place Name Types and Derivations for Sinai," located after the Preface, lists typical names and examines them in more detail.

Political Units

By the time Sinai came back to Egyptian control following the Camp David Accords of 1979, the prewar governorates were already reestablished. Two large governorates controlled most of the area, but three small areas near the canal were attached to existing governorates that have most of their territory and population outside of Sinai proper (Figure 1-2). Sinai has an area estimated at 64,500 km². Over 94 percent of the area and probably 97 percent of the population are found in the large governorates, leaving less than 6 percent of the area and 3 percent of the population attached to As Suways (Suez), Al Ismailiyah (Ismailia), and Bur Said (Port Said) governorates. Total population for Sinai was 49,769 in 1960, growing to 250,000 by 1990, a startling fivefold increase.

Sina ash-Shamaliyah (North Sinai Governorate) is headquartered at El Arish and covers 27,570 km². The north contains most of the population, 171,505 in 1989. In 1983 the population of the capital city was 56,200. By 1990 it was probably over 75,000, some 45 percent of the total.

Sina al-Janubiyah (South Sinai Governorate), headquartered at Tor, is larger in area with 33,140 km², but its 1989 population was only 28,988, 15 percent of that in the north. It is estimated that another 8,000 to 10,000 live in the area just east of the canal in the areas controlled by Port Said, Ismailia, and Suez.

Settlements

Sinai is a region of few settlements, but many camping spots, at least for the Bedouin. Foreigners are forbidden to leave the main roads. Figure 1-3 shows the thirty-one most important fixed settlements. While the list is not exhaustive, there are few other units that really deserve to be called villages or towns. In the past most settlements were small clusters of humanity around a water source which sometimes served a meager market function. Most did not exceed a few dozen families. El Arish, Rafah, and Tor have broken away from that ancient pattern.

El Arish now shows distinctive urban growth, sprawl and all. It has a large market function and is the developing center for higher education in Sinai. It has two large faculties (colleges) and more to come. Tourism is a growth industry here. It boasts a five-star hotel and numerous high-quality beach houses and condominiums.

Rafah, which spans the border with Israel, and Qantara, with a village on either side of the Suez Canal, both reflect border area

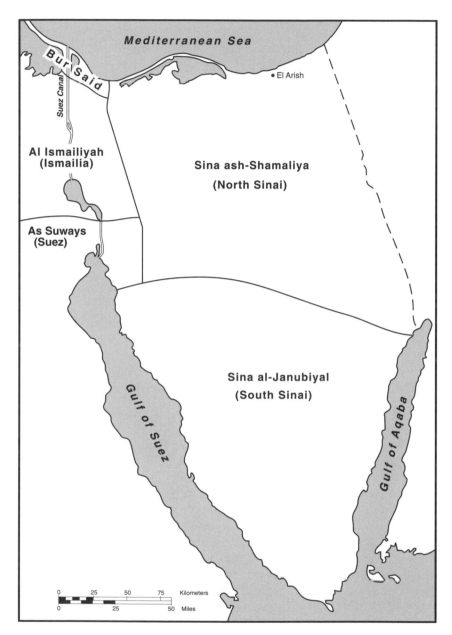

Figure 1-2. The governorates (muhafazah) of Sinai. Note the three small areas on the east bank administered by Bur Said, Ismailia, and Suez.

Figure 1-3. The thirty-one most important settlements in Sinai.

economics. A great variety of contraband can be purchased in these towns. Smuggling is among the most important economic activities in both. Border officers recently discovered an elaborate tunnel system at Rafah used to move contraband goods between Egypt and Israel. Foodstuffs, especially canned goods including cheese and butter, are prized commodities.

Qantara is very close to the free port of Port Said, and the smuggling of fabric, electronic gear, and even drugs is big business. People come from as far away as Cairo to shop because citizens cannot bring free port goods in from Port Said. Thus the smugglers perform a middleman service, at a price, but the goods are still cheaper than if a tariff had been paid.

Tor still retains a minor port function for the peninsula, but it is more important as a petroleum center and fishing town. It grows fine grapes and produces wine, but not for consumption by the Muslim citizens. It also has a long history as a spa based on hot mineral waters.

Ras Mohammed, Sharm el Sheikh, Dahab, Nuweiba, Taba, and Na'ama Bay have become important tourist towns. Scuba, wind surfing, shell and coral collecting, and other water-oriented recreation are important to their economic base. Locally, fishing provides an important income. St. Catherine and Feiran are tourist attractions for completely different reasons. They give access to mountain hiking trails and camel treks, but most importantly they have sacred and historic sites that attract a completely different breed of tourists. Higher education is also important. The Suez Canal University maintains research stations at Sharm el Sheikh and St. Catherine. St. Catherine also has a large tourist village and an airport offering access to those who would avoid the desert travel.

At the center of Sinai are settlements like Nakhl, which once served as a camping and watering place on the pilgrim road to Mecca. It now has automotive services and markets for local tribesmen. Bir Tamada, Bir Gifgafa, and Bir Hasana serve similar though lesser roles.

In selecting the thirty-one settlements it became obvious how important roads are to the present units. Places like Ain Yirga, Ain Akhdar, Ain Um Ahmed, and Ain Hudera—all well watered, but with no proper roads, only ancient caravan trails—have tended to stagnate. As a result they are losing permanent residents. To be sure, they remain significant to herdsmen who water their stock there and residents who hold titles to the land, but the importance of wage employment has attracted many of the young and progressive to other areas.

A new factor that is just beginning to affect population distribu-

Figure 1-4. A Maaza Bedouin seeding sweet melons in the sandy wadi bottom at Son el Bricki on the Wadi el Arish. Notice the funnel top on the seeding plow in which he places the seed. The draft camel is a female used for milk.

tion is Egypt's policies for resettlement of people, especially those from crowded agricultural lands in the Nile Delta. Population increase and resettlement are changing the existing cultural regionality in a variety of ways.

Cultural Regionality

The broad cultural regionality of Sinai coincides with the existing political division between the north and south. It may well have been the reason for placing the main political boundary in its east-west alignment, which follows the Darb el Haj (the ancient Pilgrim Road). Northward the tribal lands are largely held by Bedouin groups that appear to have descended from a common stock called Maaza. They claim to be Ashraf, direct descendants of Mohammed through his daughter Fatima. Prior to 1768 elements of the Maaza were found all over Egypt, but policies of extermination by the Mameluke rulers left the main remnants in Sinai and Egypt's Eastern Desert. Today the Sinai remnants pasture the sandy plains of the Mediterranean Littoral and the northern part of the Tih Plateau. They practice horticulture and fruit husbandry around wells and in favorable wadi bottoms (Figure 1-4).

South of the Pilgrim Road, a shifting confederation called the Towara dominates the economic landscape. The core of this group came en masse from eastern Egypt in the seventh century. Another important component originated in the Hejaz region of the Arabian Peninsula. It is uncertain when they first came to Sinai, but they were already a dominant tribe when Mohammed Ali came to power in 1805. The Towara annihilated or absorbed older Bedouin groups then living in the Sinai Massif, the Qa Plain, and the Aqaba Foreshore. Only small groups exist outside the confederation. Their economic and cultural impact is indeed minor.

On balance the tribes of north Sinai are stronger than those of the south. In both regions internal feuding has been a way of life, but today modern economic realities and urbanization are blurring the distinctions of the past. The Bedouin of Sinai are hillbillies to Egyptian Arabs, but the Fellahin still fear these fierce men of the desert, who are only now succumbing to the enticements of economic consumerism.

On the other side of the coin, Egypt tends to view Sinai as a relatively open land able to sustain more population and geopolitically needing Arab citizens to secure it against Jewish occupation. The success of reclamation projects and new settlements will largely depend upon how Egypt perceives the political needs and how vigorously it pushes economic development. The final chapter will consider some of these problems in view of sustainability as well as political reality. It is far from certain that Sinai can absorb vast numbers of people, whether they come because of migration policies or result from natural increase of native population. It is a great land, but a fragile land.

2. Plate Tectonics and the Geology of Sinai

Sinai is truly a land which flaunts its geologic and tectonic past. Many desert areas are as denuded of vegetative masking, but few have the complexity of their origins as well delineated as Sinai. It is a wonderland for the physical geographer, the historical geologist, the geomorphologist, and the student of plate tectonics. Donald Katz depicts the area as a landscape "still ringing in the afterglow of a geological trauma some 6 million years old that folded it until it snapped; southern Sinai is old, but not the least bit elegant in its old age." The trauma is certainly evident, but Katz is mistaken— for all its austerity Sinai is really quite elegant.

Plate Tectonics

Plate tectonics provide us with recent information about the oldest formative processes in the region, which also helps explain Sinai's location between Africa and Asia. It is this pivotal position that enhances its strategic prominence in both the physical and cultural geography of the Near East. The present landmass is tied to both the African and Asian continents, but its above-sea-level link with the Arabian plate is now more continuous than the corresponding tie with Africa. Yet geologically, the peninsula is perhaps a bit more closely related to the African mass. This is especially true of south Sinai, where the ancient crystalline mountains are an extension of the Red Sea Mountains of the African mainland. The reddish hue of these mountains gives name to the "Red Sea" rather than the water, which tends to be very blue, reflecting the usually cloudless skies.

To summarize the more obvious movements of crustal materials up to this point in geologic time, it is necessary to begin 800 million years ago, with the oldest rocks known in Sinai and throughout the Levantine Plate on which Sinai sits. These ancient metamorphics, formed from sediments mostly of the clay and silt size ranges, are found in the regions of Aqaba and Wadi Feiran. Above the metamorphics is a thick prism of sedimentary rocks deposited during a long period of passive activity.

This low-energy period was followed by continental collision. Massive folding and metamorphism of the sedimentary materials turned them into schists formed by smearing out under heat and pressure. In the hotter and deeper areas of deformation, partial melting created "migmatites" (mixed rocks) and allowed the intrusion of readily mobile elements, such as alkali metals, into the molten mass of the older rocks. Complexity was increased again by intrusions of basic gabbro and less basic tonalite magmas into the crust. Plutons of granite, granodiorite, and quartz-diorite, derived from slow-cooling magmas, were intruded through these older materials as east-west-oriented dikes, which cooled beneath the surface. About 700 mya (million years ago) rocks of the Kid and Sa'al groups were eroded from the emergent landmass and deposited in a marine environment made up of Elat and Feiran materials. Subduction is a type of plate collision where an oceanic plate moves under a lighter continental plate. Such a subduction triggered volcanic activity along the continental margin in what is now south Sinai. Ancient lava flows and sediment accumulation in the Wadi Kid area resulted in layers up to 10 km thick.

As with most areas of the world, the earliest events of Sinai's geologic history are disputable. The origins of its basement rocks remain in question. Geologic evidence seems to indicate that they were part of an old continental crust over which younger ocean sediments accumulated and lithified. This development of the Sinai basement is tied to events that created the larger Arabo-Nubian Massif, which rose from a primal ocean centered over what is now Arabia.

Batholithic Intrusion

The intrusion of plutonic materials was nothing new, but during the late Precambrian (620 to 580 mya) there was an unprecedented thrust of granitic plutons into the Arabo-Nubian area. Specifically, these batholithic intrusions formed the present Sinai Massif, with its rugged mountains such as the St. Catherine, Serbal, and Um Shomer groups. Later volcanism (Cambrian—580–540 mya) produced rather interesting features called "ring-dikes," which were formed by the intrusion of syenite into areas weakened by the collapse of ancient calderas.

A long, erosive period followed the final Precambrian uplift and resulted in the Infra-Cambrian Peneplain. Upon this plain were deposited the typical Lower Cambrian materials of Sinai, arkosic and fluviatile conglomerates. Erosion was so thorough, however, that the only known outcrops of this type are found in the area between Taba and Timna northeast of the Sinai Massif and at Um Bogma on the west. Near the end of Cambrian times the sea retreated northward, exposing sedimentary sandstones, silts, and clays. Other materials of the lower Paleozoic are largely missing from the Sinai geologic column. However, prior to the formation of the supercontinent, "Pangaea," by the merger of Gondwanaland and Euramerica during the Silurian Period (350 to 320 mya), Sinai and the Arabian segment were locked solidly into the African Crustal Plate.

A great erosional unconformity removed much of the record of what had occurred during the lower Paleozoic. The little evidence there is suggests that for most of Cambrian, Ordovician, Silurian, and Devonian times, Sinai was a continental shelf receiving a mix of marine and nonmarine sediments. During the Carboniferous, southeast Sinai was the only emergent portion.

Mesozoic

Beginning with lower Triassic (255 mya) and extending to the Eocene (58 mya), the shallow Tethys Sea, with its broad continental shelf, was the dominant feature over north Sinai. Paleogeography of the Jurassic (210–145 mya) would show the emergent part of Sinai being southwest of a line running from Port Said to Taba. North and east was a shallow continental shelf adjoining the Tethys Geosyncline (Figure 2-1).

At the beginning of lower Cretaceous (145 mya), the Levantine Crustal Plate, on which Sinai sits, was still locked into the larger African Plate. The Tethys Sea separated the Afro-Arabian area from the Eurasian landmass. Africa then included, in one crustal unit, the mass that would eventually pull apart into the Levantine, Anatolian, Iranian, and Arabian plates (Figure 2-1). Lower Cretaceous was marked by an extensive regional uplift. The thrust was, however, uneven, shoving southeastern Sinai more rapidly upward, thus tilting previously deposited materials an average of 2 degrees downward toward the northwest. The uneven uplift and subsequent erosion caused greater denudation in the south, where over 1,000 m of sediments were removed, compared with less than 100 m in the north. By middle Cretaceous (Cenomanian), all of Sinai was under a shallow sea. During lower Turonian, Sinai was subjected to broad folding, resulting in structures extending over hundreds of

Table 2-1. The Geologic Column of Sinai

ERA	SYSTEM	EPOCH		TIME m.y.a.	EVENTS IN SINAI	WORLDWIDE EVENTS
Cenozoic	Quaternary	Pleistocene		2	Sinking of N.W. Sinai Dune sheet accumulating	Continental Glaciation Advent of Homo Sapiens
Cenozoic	Tertiary	Pliocene		6	N.W. half covered with Proto-Mediterranean Sea [1]	Dominance of Elephants, Horses, Large Carnivores
Cenozoic	Tertiary	Miocene		22	West 2/3 covered by Tethys Sea [1]	Development of Monkeys and Whales
Cenozoic	Tertiary	Oligocene		36	Most of N. Sinai in Tethys Sea Geosyncline	Rise of Anthropods
Cenozoic	Tertiary	Eocene		58	Little Fossil Record [1]	Grasses
Cenozoic	Tertiary	Paleocene		63	[1]	Primitive Horses
Mesozoic	Cretaceous	Senonian		88		Flowering Plants
Mesozoic	Cretaceous	Gallic	Turonian	90		Spread of Mammals
Mesozoic	Cretaceous	Gallic	Cenomanian	97		Extinction of Dinosaurs
Mesozoic	Cretaceous	Neocomian		145	Tethys Geosyncline Developing	

Table 2-1 (continued)

ERA	SYSTEM	EPOCH	TIME m.y.a.	EVENTS IN SINAI	WORLDWIDE EVENTS
Mesozoic	Jurassic		210	[2]	Birds
	Triassic		255		Conifers Spread / Dinosaurs
Paleozoic	Permian		280	Most of Sinai a part of Gondwana	Conifers Abundant
	Upper Carboniferous		320	S.E. Sinai emergent, Marine deposits represent this time in other parts [1]	Insects, Reptiles / Coal Forests
	Lower Carboniferous		360		Amphibians Widespread
	Devonian		415	[1]	First Amphibians
	Silurian		465		Land Plants / Primitive Fish Life only in seas
	Ordovician		520		
	Cambrian		580		Trilobites / Marine Invertebrates
		Precambrian		[3]	Green Algae / Bacteria, Blue-Green Algae

Notes:

[1] Geologic strata missing or poorly represented in Sinai

[2] Jurassic materials are found in north Sinai only; in central Sinai, poorly correlated Nubian Sandstone is found in the geologic column where Jurassic, Triassic, and Permian materials might be expected

[3] The Precambrian rocks of the southern tip of Sinai were part of the Ethiopian Shield, probably the most stable block extending foward into the Phanerozoic

Source: Siever (1983), Harland (1989), and Kummel (1970)

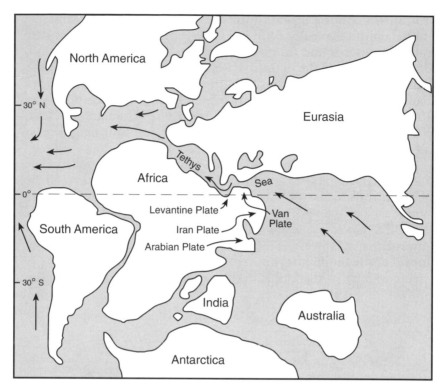

Figure 2-1. Cretaceous sea-land arrangements. During the Cretaceous Period (100 mya) narrow sea basins, following the Tethys Geosyncline, separated the African-Arabian area from the Eurasian landmass. The Indian Subcontinent was adjacent to east Africa (not yet in collision with Asia). The Levantine and Arabian plate fragments had not begun to separate from the African plate. Adapted from Brian Windley (1977), A. G. Smith (1977), and Samuel Mathews (1967).

km and up to 1,000 m high. The most prominent of these structures is the fold extending from Gebel Maghara westward past Mitla Pass.

Forming the Peninsula

Up to middle Cretaceous, Sinai had no distinct spatial configuration to separate it from the rest of the Arabo-Nubian Massif resting on the greater African Plate. Beginning in the late Mesozoic (Senonian–Cretaceous) new forces were at work, poised to tear apart the massive African Plate. The tectonics which shaped peninsular Sinai as a distinct entity started with plumes of deep-seated volcanism doming upward into the continental mass. Numerous domal uplifts, typically 1 km thick, 100 km wide, and 300 km long, radically altered and thinned the crust, allowing the formation of deep rifts such as the extensive Red Sea, East African, and Gulf of Aden rifts (Figure 2-2). By the Miocene Epoch, these massive rifts had coalesced to form the Afar Triple Juncture. The Gulf of Aden Rift is an extension of the fracture known as the Indian Mid-Ocean Ridge, which thrust westward into the African continental mass, where it intersected the Red Sea and East African rifts.

During early Miocene (22 mya), continued subsidence of the thinned continental crust allowed the Red Sea Rift to extend farther northward as the Gulf of Suez, creating a definitive plate boundary between Africa and the newly forming Levantine Plate. The Sinai block was again subjected to uplift with a downward tilt toward the northwest. This rejuvenation caused considerable removal of the sedimentary cover. Magmatic intrusions, 21–19 mya, created many of the basalt dikes so prominent in Sinai's present surface geology and geomorphology (Figure 3-9).

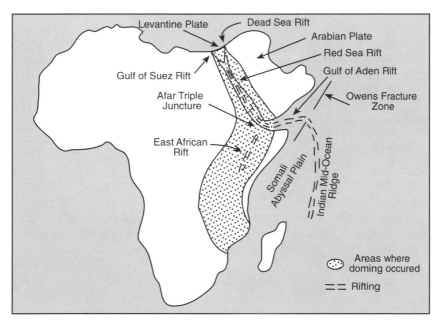

Figure 2-2. Domal uplift and rifting in the ancient Afro-Arabian Plate. Adapted from Brian Windley (1977), Samuel Mathews (1967), and Rolf Meissner (1986). Map based on "Pilot Chart of the Indian Ocean," U.S. Hydrographic Office, and National Geographic Map of Indian Ocean Floor, October 1967.

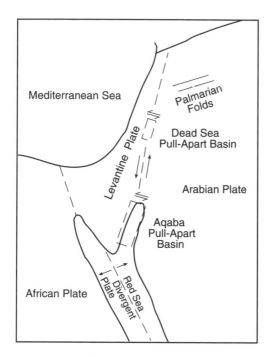

Figure 2-3. The divergent boundary between the Arabian and Levantine plates.
Note the "pull-apart" basins of the Dead
Sea and Gulf of Aqaba. Relative displace-
ment amounts to 105 km in the south, but
much less in the north, where convergence
with the Van plate is slowing the progression. Adapted from Clark Burchfield (1983).

By the Pliocene (3.5 mya), the thinning of the original continental plate materials had progressed to the point that ocean crust was
forming in the southern portion of the Red Sea. In the northern
part of the sea these features remain obscured by a thin continental crust.

At the same time as the southern Red Sea was developing the
divergent boundaries characteristic of an ocean crust, the Dead Sea
Rift was being extended in a northeasterly direction, driving a
wedge between the Levantine and Arabian plate fragments (Figure 2-3). The Dead Sea Rift forms a north-south interconnect between the Red Sea divergent plate boundary and a convergent plate
boundary at the Taurus Mountains in Turkey. Along this boundary between the Arabian and Levantine plates, the north-south-
oriented Dead Sea Rift is offset in some places to the west, forming

a series of "pull-apart" basins including the Gulf of Aqaba and the Dead Sea itself. Within the immediate vicinity of Sinai, strong divergence (east-west pull apart) and convergence (north-south compression) are indicative of the dynamic tectonics that are in the process of tearing the Levantine Plate away from Africa and grinding the Arabian Plate past the Levantine segment as it is shoved into the Asian landmass. Thus Sinai, which makes up much of the Levantine Plate, continues to change its relationship with both Africa and Arabia.

The northward movement of the Arabian Plate into the Eurasia Mass has resulted in the absorption and shortening of this bit of crust. The closure between the Arabian and Anatolian plates is evidenced by the Palmarian Folds in Syria, formed by the compressive action of convergence. This type of tectonics thickens the continental crust by adding new materials from the mantle.

Since Pliocene the geologic history of Sinai has reflected a renewed uplift of the central and southern regions. Marine sedimentation has dominated events north of Gebel Maghara, where clastic sediments from the Nile River have been added to the coastal margins. The Nile has been flowing in its present valley since the end of the Eocene. The occurrence of recent evaporites is related to climatic desiccation and a shallowing of the Mediterranean.

From Taba to Ras Mohammed, fault cliffs are indicative of recent and continued tectonic uplift. From Nuweiba northward there has been salt deposition in shallow coastal lagoons, which were formed by tectonic changes. Modern Sinai is truly the gift of restless continents, but instead of being characterized as the "meeting place of continents," it should be thought of as the incompletely severed umbilical cord still attaching the breakaway Arabian Plate to Mother Africa.

Surface Geology

In spite of difficult conditions imposed by rugged and desolate terrain, shifting sands, lack of access roads, and antagonistic Bedouins, a surprisingly detailed geology of Sinai was developed by the end of the first quarter of the twentieth century. The works of T. Barron (1907), W. F. Hume (1906), F. W. Moon and H. Sadek (1921), and H. J. L. Beadnell (1927) provide us with excellent background materials. For all our recent efforts in remote sensing and GIS (Geographic Information Systems), much of the current ground truth still depends on the work of these hardy, camel-riding pioneers of Sinai scientific exploration.

The broad structure of Sinai geology falls easily into two parts:

the Precambrian base that is widely exposed in the south, and a wedge of sedimentary layers in the north that generally becomes thicker and more recent toward the Mediterranean coast. The boundary between the two runs west to east from Gebel Hamman Faraun on the Gulf of Suez to Nuweiba on the Gulf of Aqaba, and coincides with the Tih and Egma escarpments in the high central part of the peninsula. On the basis of age, the surface geology can be further subdivided into seven groups (Figure 2-4). From oldest to most recent these are: Precambrian intrusives and metamorphics, Jurassic sedimentary strata, Cenomanian–Turonian limestones and dolomites, Senonian chalk, Eocene chalk and limestones, Oligocene and Miocene sediments and Miocene dikes, and Quaternary alluvium. (See the geological time frame in Table 2-1.)

The south is dominated by the Sinai Massif, which is composed of ancient Precambrian rocks (Figure 2-4). An estimated 75 percent of this mountainous area is made up of intrusive crystalline rocks. The other 25 percent is metamorphic in origin and predates the crystalline materials.

The crystalline areas were created by the intrusion of at least sixty plutons. Eons of geologic time have since removed essentially all crustal overburden in which the intrusions occurred, leaving the Precambrian surface largely unadorned with more recent materials. The areal extent of individual plutons ranges from 50 to 200 km², and together they make up the dominant surface.

The metamorphic materials are scattered through the dominant crystalline rocks, but tend to be concentrated in four areas:

1. Taba-Nuweiba—in the east, a narrow belt of schists, gneisses, amphibolites (crystallines dominated by hornblende), and migmatites (granite-like intrusives) that parallels the Aqaba coast.
2. Wadi Sa'al—on the northeast of the massif, made up of metasedimentaries and meta-volcanics, most of which were modified from conglomerates and pyroclastics.
3. Wadi Kid—in the southeast, similar in composition to the Sa'al materials.
4. Wadi Feiran—northwest of the massif, made up of rocks similar to the Taba-Nuweiba concentration, but with the addition of quartzite and marble. This last group is also found in smaller exposures west of Dahab, along Wadi Sulaf and at Gebel Catherine.

Dikes resulting from the intrusion of basaltic magmas into the fissures and weak zones of overlying formations are common.

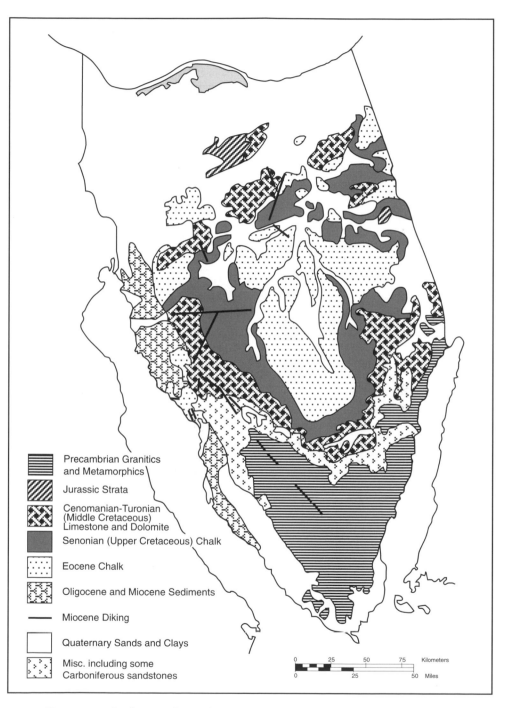

Figure 2-4. Surface geology of Sinai.

Precambrian Granitics
and Metamorphics

Jurassic Strata

Cenomanian-Turonian
(Middle Cretaceous)
Limestone and Dolomite

Senonian (Upper Cretaceous) Chalk

Eocene Chalk

Oligocene and Miocene Sediments

Miocene Diking

Quaternary Sands and Clays

Misc. including some
Carboniferous sandstones

0   25   50   75   Kilometers

0   25   50   Miles

Diking is perhaps Sinai's most singular geologic signature. Dikes form a significant part of the intrusive materials found in the older Precambrian rocks exposed in the massif (Miocene volcanic intrusions are discussed below).

Before 1912 it was thought that no rocks of Jurassic age were to be found in Sinai. The discovery of such strata in the Gebel Maghara led to exploratory drilling and the hopes of discovering underlying carboniferous materials. The oil industry benefited very little from this activity, but the geology of north Sinai became better known.

The Jurassic (Lias) rocks of Gebel Maghara and several smaller mountains reflect a series of marine transgressions and regressions. These movements left behind marine sediments of limestone, marls, and clays. Later tectonic activity folded, faulted, and metamorphosed these deposits, greatly confusing their stratigraphy.

In central Sinai, Nubian Sandstones (Figure 2-5) are usually found in the geologic column where Jurassic, Triassic, and Permian materials would normally occur. The broad term "Nubian Sandstones" was applied as early as 1837, and subsequent use has not always been consistent. By 1927, Beadnell and others had suggested that the term was more accurately applied to petrology than to a measure of true stratigraphical layering.

There is little evidence of Lower Cretaceous deposition in Sinai; however, Middle Cretaceous (Cenomanian) and early Upper Cretaceous (Turonian) lime materials are well represented. In Figure 2-4 it can be seen that the sedimentaries of the central plateau are surrounded by nearly continuous exposures of limestone and dolomites of these periods. Technically, central Sinai has rocks ranging in age from Cambrian to Neogene (late Tertiary), but the bulk of the surface material is dominated by three concentric and increasingly younger bands: (1) Cenomanian–Turonian limestone and dolomite, (2) Senonian chalk, and (3) Eocene chalk (Figure 2-4).

The Senonian Epoch sits in the highest part of the upper Cretaceous System. The name "Senonian," like "Cenomanian" and "Turonian," is more commonly used in Europe and the Near East than in North America, but there are no inconsistencies as far as the geologic column is concerned. During the Senonian, northern Sinai was under deep ocean water, as part of the geosyncline (broad downwarp of crustal materials) occupied by the Tethys Sea. Shoreward from the deep water, a wide area of transgression, caused by crustal warping, extended the continental shelf area deep into central and south Sinai. The distribution of Senonian chalk (Figure 2-4) reflects sedimentation onto this shallow shelf area.

Exposures of Eocene chalk are dominant in gebels Tih and Egma.

Figure 2-5. Honeycomb, Nubian Sandstone, Wadi Zaghra east of Gebel Musa.

Gebel Egma actually sits astride and forms the highest part of the Tih Plateau, with the Egma Escarpment delineating the southern boundary of the Tih as well as the southernmost exposure of Eocene materials. North of Gebel Tih are several smaller exposures, including those surrounding Gebel Raha.

Lower Eocene materials, called "Libyan," are represented by Egma Limestone and its associated underlying marls and shales. Beadnell called the middle Eocene rocks "Mokattam" (probably for the upland area east of the Cairo Citadel). In this group are included chalky limestones and cardita beds underlain by nummulitic (containing marine foraminifera) limestones. Upper Eocene in Sinai includes red beds, gypseous marls, and green marls. Due to the general domal configuration of central Sinai, Eocene chalk is more centralized than Cretaceous materials.

The stratigraphic geology of the Oligocene in Sinai is quite unimportant. Beadnell (1927) doubted it was represented, but according to the *Atlas of Israel* (1985) it is found in association with Miocene clays, sandstones, marl, conglomerates, and limestones interbedded with gypsum. These scattered occurrences are limited to the Gulf of Suez side.

Miocene volcanic intrusions have produced a variety of dikes, sills, and other features made of olivine basalt and diorite. As noted earlier Miocene diking is in many ways the most striking geologic phenomenon in Sinai. These intrusions appear in every geologic formation not covered with Quaternary sediments. Overall they make up nearly 5 percent of the total surface, but in some areas the density is great enough to comprise more than 50 percent of the surface mass. These landforms include the interesting "ring-dikes," so called because they were intruded around the margins of ancient calderas, particularly in the Catherine area. Some of the larger and more prominent dikes are mapped in Figure 2-4, but the multitude of these features is far too great to map on anything but the largest-scale geologic maps.

Purportedly the longest dike system in Sinai is the slanted "T-shaped" figure shown in Figure 2-4. Beadnell indicated that the offshoot of this great dike running southwest from Gebel Bodhia to Gebel Um Retama covers a distance of 28 miles (44.8 km). He further implied that it probably continues another 12 miles (19.2 km) to the coast near Gebel Hamman Faraun.

Recent geologic surfaces of Sinai are made up of aeolian sands, wadi alluvium, clays, conglomerates and gravel terraces, and uplifted beaches and coral reefs. One hesitates to use the word "Pleistocene" in conjunction with these materials, as there is a tendency to regard the Pleistocene as an epoch of the Cenozoic distinguished by the occurrence of glaciation. Since no Pleistocene glaciation is known to have occurred in Sinai, this recent geologic time frame will be simply referred to by the system name "Quaternary."

While glaciers gripped extensive land surfaces to the north, Sinai received a more gentle climatic change in the form of increased precipitation and greater fluvial erosion. Quaternary materials dominate the northern plain areas. Here vast dune sheets cover much of the surface. Elsewhere wadi channels and narrow coastal plains have exposed clays, conglomerates, and marls. Similar exposures sometimes reveal small inclusions of older rocks, particularly those of Cretaceous age, at gebels Yelleq and Halal. Traces of Jurassic materials are found with Quaternary materials at Gebel Maghara.

In the northwest, the present-day shoreline of the Mediterranean is associated with a tract of sandy dune ridges that sometimes exceed 300 feet (91.4 m) in elevation, but the nearby salt marshes of the Tenna Plain are sinking because of gentle folding along the edge of the Mediterranean. Prior to construction of the Aswan High Dam, this downwarp was compensated for by the eastward drift of Nile sediments. Now the whole delta area has shifted from a prograde to a retrograde pattern of deposition. It yet remains to be determined what the long-range effect of shoreline erosion will be upon the Tenna Plain and the adjoining Sabkhet el Bardawil, with its important fishing industry.

In the central plateau area, Quaternary materials are found in the wadi valleys. Other wadis that rise in the uplands of Yelleq, Maghara, and Halal simply sink into the sands of the dune sheet and lose their identities as drainage channels.

In the south, Quaternary materials also cover the Plain of Qa, which fronts on the Gulf of Suez. Like the Mediterranean Littoral, these are recent sediments, mostly fine sand and gravels eroded from the Sinai Massif to the east. In the fault-controlled environment of the Red Sea, south Sinai is strongly influenced by a divergent plate boundary. The pull-apart basin forming the Gulf of Aqaba is more active than the Gulf of Suez (Figure 2-3), resulting in the formation of a broader shoreface plain and a greater accumulation of recent alluvium on the Suez side. Only the southernmost sector of the Aqaba shoreline has enough tectonic stability for coastal plain development and the accumulation of Quaternary deposits.

In summary it may be said that the role of plate tectonics is greatly evident in the geologic column of Sinai. In the south the ancient massif was formed from a complex of intrusive volcanic and metamorphic materials associated with plate movements and adjustments between crust and mantle. In the north the expansion and contraction of the ancient Tethys Sea controlled the deposition of the dominant sediment cover. Through most of this depositional period the Tethys completely separated Eurasia from the Afro-Arabian Plate. The deposition of sediments has carried over into the recent geologic past, when the Nile River continued to dump sediments along the increasingly shallow end of the eastern Mediterranean Sea, which formed a portion of the ancient Tethys, but is now closed by the abutment of the Arabian Plate against Asia. The total geology is one of great intricacy and richness that contributes to the mineral wealth of modern Sinai.

3. Geomorphology and Drainage

As in most desert environments, the geomorphology of Sinai is dominated by erosional features usually caused by water. Even the shifting sands of the northern dune sheet show some effects of water movement, but in the great Tih Plateau and the Sinai Massif the imprint of running water dominates the landscape.

In this chapter the relationship of geomorphology to drainage will be examined for each of nine regions. For some units like the Dune Sheet and the Suez Foreshore the discussion will be a rather straightforward examination of physical characteristics. For others such as the Sinai Massif the dominance of places like Gebel Musa and Gebel Serbal has such historic significance that it is nearly impossible to discuss landform features without relating them to historic or legendary events—their names alone would preclude any other approach.

Northern Sinai is dominated by undulating, aeolian sand plains beginning at the Mediterranean Sea and gradually rising southward toward dissected plateaus of Cretaceous and Tertiary sediments. A profusion of unconsolidated and aeolian reworked materials of Quaternary age dominates the northern dunes and sand plains. In the southern part of the Dune Sheet a few outliers of massive limestone, such as gebels Yelleq, Maghara, and Halal, rise above the undulating surface, which is mostly erosional debris of soft, chalky rubble. In central Sinai the Tih Plateau continues the dominant southward rise, culminating at the Tih and Egma escarpments with elevations as high as 1,600 m. South of the escarpments elevations drop where the drainage systems of Feiran to the west and Watir to the east cut their way between the plateau and the granitic Sinai Massif to the south. Figure 3-1 maps the nine geomorphic regions of Sinai. Figure 3-2 provides a vertical cross-section of the north-south profile.

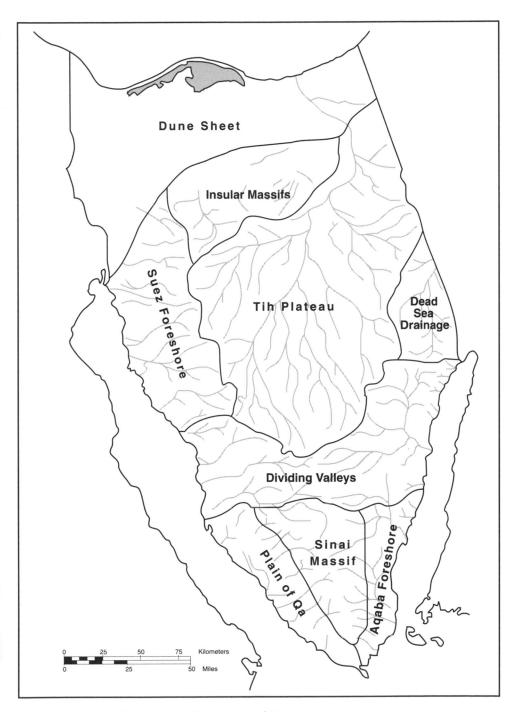

Figure 3-1. The geomorphic regions of Sinai.

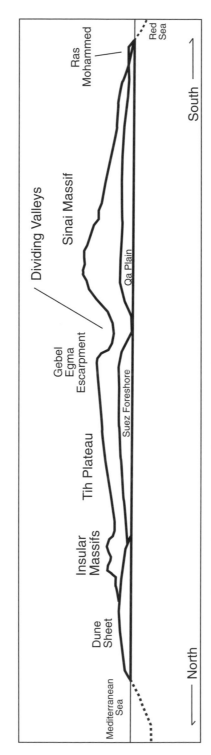

Figure 3-2. The north-south, vertical cross-section of Sinai geomorphology as viewed from the west side.

The Dune Sheet

North Sinai is dominated by a great crescent of loose sands. From its southwestern horn, southeast of the Port of Suez, the Dune Sheet arcs northward along the margin of the Suez Canal, then eastward along the edge of the Mediterranean all the way into Israel, tapering out near Beersheba. At first glance it would be easy to mistake the dominance of sand and low color contrasts for morphological uniformity. In spite of nearly constant aeolian movement in many sectors, the Dune Sheet hides a surprising variety of relatively stable habitats that support small communities of life. The influence of an ain (spring), a bir (well), or themila (area of shallow groundwater) can be seen in many places. It comes almost as a shock for a traveler, crossing high-profile aeolian dunes, to suddenly drop into a hidden valley where ancient palm or sidri trees are sequestered behind nearly vertical walls of seemingly unstable sand.

The great crescent may be divided between an inner and outer band. The most distinctive feature of the outer band is the vast, shallow lagoon, Sabkhet el Bardawil. It was once differentiated from the Mediterranean by an unbroken barrier island. Now man-made openings through the beach ridge connect the saline waters of the lagoon with the Mediterranean proper.

The inner band of the crescent begins at Mitla Pass, growing broader to the northwest of Khatmia Pass, where it curves eastward past Gebel Maghara. In space photos, Gebel Maghara appears as a great, dark eye in the broad crescent of light-colored sand. Farther east the Dune Sheet shows its maximum relief with individual dunes sometimes having more than 50 m of vertical rise.

Between El Arish and Rafah the dunes tend to form stable hillocks where brush clings to the north slopes, which have lower rates of evapotranspiration. South of Rafah, distinctive dune forms provide an interesting topographic variation. Flat sand plains are often completely surrounded by high dunes, giving the appearance of a pit—"al juwra" in the Bedouin equivalent.

Insular Massifs

Three limestone massifs rise above the Dune Sheet of north Sinai. Gebel Maghara is completely surrounded by sand, but Gebel Yelleq and Gebel Halal form the southern margin, with dunes lapping on their northern flanks.

The cores of these ancient ranges are folded, creating domal structures into which erosional valleys have been cut. Yelleq is the highest and most prominent. Its highest peak, Ras Guran, is 1,094 m.

Figure 3-3. Cuesta formed in sedimentary beds tilted by the doming of Gebel Halal.

This point aligns with other peaks to form a northeast-southwest-oriented backbone. The southern wadis are very steep and many contain sidds (dry waterfalls). All wadis soon lose their identity as they merge with the Dune Sheet to the north or empty onto the broad drainage of Wadi Bruk in the south.

Gebel Maghara has a large erosional basin cut into its dome. Here the Shusht (summit) Maghara is accented by concentric escarpments. Short wadis move water internally, but there is essentially no surface drainage onto the Dune Sheet, which laps high upon its limestone flanks.

East of Yelleq is Gebel Halal. Its geomorphology is similar to that of Yelleq, but on a smaller scale. The doming of Halal and subsequent erosion have created cuesta landforms around its flanks (Figure 3-3).

Suez Foreshore

The Suez Foreshore encompasses the low-lying plains from Wadi Giddi to Wadi Gharandal. The foreshore plain extends from sea level to approximately 250 m in a series of low-profile terraces. Near the gulf, irregular gravel terraces dominate the environment. In places the gravels grade toward coarse conglomerates, but the lowest areas are sometimes covered with recent fine sand. The monotony of this surface is occasionally broken by widely scattered wadi channels and a few low-lying hills.

East of the terrace area are plateau remnants of Eocene limestones. Between Gebel Raha and Mitla Pass, extending into the upland areas of Wadi Haj, is an elevated tableland bounded by vertical cliffs and prominent headlands like Raz el Jeifi.

The western side of these uplands overlooks an area covered with desert scrub, which gives way to a veritable sea of sand. The eastern side looks across the gentle topography of the Tih Plateau. Between the peaks, rugged gorges, which F. W. Moon characterized as "cul-de-sac-wadis," have been cut into the edge of the plateau.

Tih Plateau

The Tih Plateau occupies the central part of Sinai. It is sometimes called the "Desert of the Exodus," or the "Wilderness of the Wandering," because it is thought that much of the wandering of the Children of Israel occurred here. Tih means wandering. Emptiness and absence of contrast are the overwhelming impressions left with travelers who cross its broad expanses. The modern traveler would have to get well away from the main roads to find any real contrasts or exciting views, and there are few side roads to provide for such. Most of the desert side roads in Sinai are off-limits to foreigners because of the danger of becoming stranded in hostile environments.

From its highest sectors in the south, the elevated limestone plateau slopes down gently northward along the natural dip of the crustal plate. In the north the plateau terminates against the Insular Massifs. The southern and western edges are well delineated by extensive escarpments. In the east toward the edge of the Dead Sea drainage the plateau breaks into scattered peaks. To the south the high escarpment of Gebel Egma separates the plateau from the Dividing Valleys.

Together the escarpments of Tih and Egma have always impeded north-south travel, focusing it on narrow passes like Naqb Rakna and Naqb Mirad, which became ideal spots for Bedouin plunder (see Dividing Valleys below).

The southern half of the plateau is the most desolate part of Sinai. The barren, stony (reg) plateau is exposed to intense solar radiation, which, combined with scant precipitation, precludes almost all plant growth. Rough, stony surfaces make it difficult to cross.

The El Arish Basin covers a total of 13,720 km², about 28 percent of Sinai. The Tih Plateau, that portion of the basin above the Halal Narrows, covers 12,740 km². Elevations range from 1,600 m

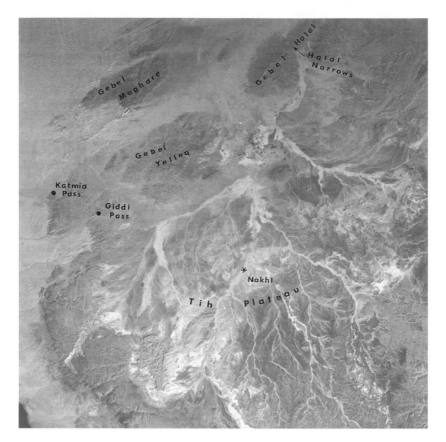

Figure 3-4. Satellite imagery of the Tih Plateau. Note the bright color of the sand valleys compared to the dark patina of the hammadas in the interfluve areas (NASA ERTS, November 9, 1972).

in the south down to 150 m where Wadi el Arish cuts through the narrows. The wadi trajectories are parallel and flow in a consistent northward pattern, reflecting a long, uniform slope.

This vast surface is dominated by a dark gravel plain patinated by desert varnish on a relatively stable surface of gravels and exposed bedrock. The high chromas of the sand surfaces stand in stark contrast to the low chromas of the desert varnish (Figure 3-4). In some areas flint is a widespread, but relatively minor, component of the surface. Smaller flints sometimes occur in the chalky gravels.

Near Nakhl, the slopes of the Tih become very gentle, almost flat, with elevations dropping gradually from 500 m to 150 m where the wadi exits through the Halal Narrows (called simply "Adh-Dhayga"—The Narrows—by the Bedouin). In this sector, the wadis increasingly converge and the whole system is forced toward the northeast by the massifs of Giddi and Yelleq. Extensive sand deposits clog these broad valleys. Drainage channels tend to wander aimlessly over and into these flat sand areas designated here as "sand valleys," to differentiate them from the Dune Sheet. This drainage phenomenon is sometimes referred to as "unchanneled flow."

North of Nakhl, sand valleys form the dominant cover, and in places the branching drainages become absolutely confused. The sand valley formed by the coalescence of Wadi el Arish and Wadi Geraia has an unconsolidated surface 10 km across, but downstream it is constricted to 500 m where it passes through the Mitmetni-Bedan Narrows (Figure 3-5). The whole system then cuts through the eastern end of Gebel Halal and exits from the Tih Plateau to fight its way across the Dune Sheet to the Mediterranean Sea.

Dividing Valleys

Geomorphologically, the mountainous south has been traditionally delineated from the central plateau along the Egma Escarpment. In reality the Sinai Massif does not abut directly against the escarpment. Instead, they are separated by a system of valleys 20 to 30 km across. These valleys provide a definitive break between the chalky plateau to the north and the granitic massif to the south (Figure 3-6). The morphologically unifying factor is the rugged landforms cut into sandstone. Wadis like Garf and Zelega have eroded through the original overburden to expose the only significant expanses of sandstone in Sinai.

Near the point where Wadi Garf enters Wadi Baba the sandstones give way to marine deposits of Carboniferous age. Here are found the ancient workings for turquoise at Serabit el Khadim and the manganese mines of Um Bogma.

The rich oasis of Wadi Feiran is the best-known feature in the Dividing Valleys. Above the oasis the channel is a dry torrent bed, impinged by canyon walls. The bare rock slopes are bold, rugged, and highly chromatic, with gneiss, mica-schists, greenstone, and granite providing a wide variety of color and texture. Wadi Feiran has no sandstone in its middle or lower reaches. After leaving the granitics it empties onto a broad, sandy plain, eventually becoming lost behind Ras Sharatib, without reaching the sea. East of the divide, the geologic structure cuts across the general erosional slope.

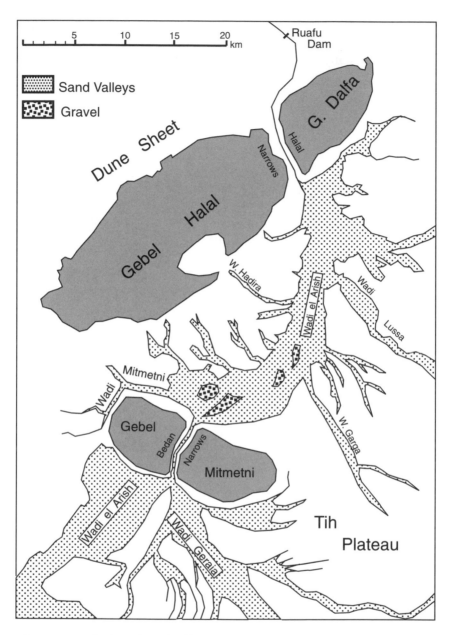

Figure 3-5. The broad sand valleys of the Wadi el Arish and its tributaries. Above each of the "narrows" the flat surface of the wadis widens out to 8 or 10 km.

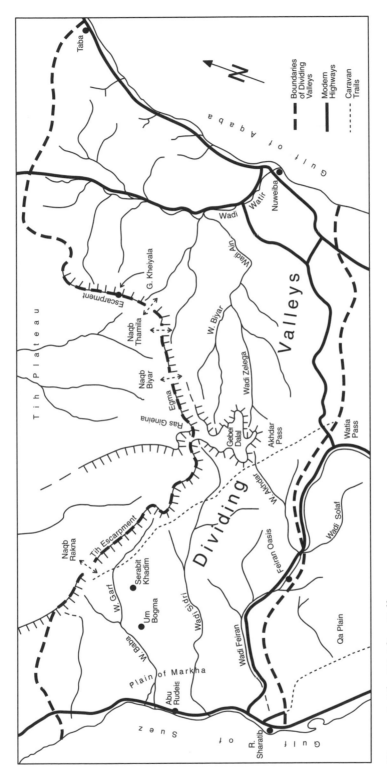

Figure 3-6. The Dividing Valleys.

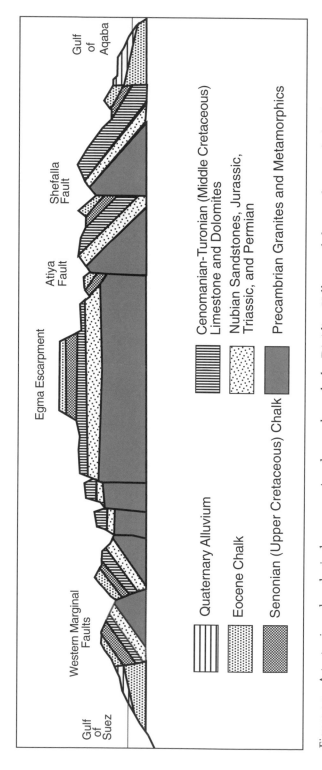

Figure 3-7. A tectonic and geological cross-section drawn through the Dividing Valleys and the southern end of Gebel Egma (adapted from Beadnell, p. 23).

This conflict of slope and structure creates a jagged pattern in the deep-valleyed wadis. The wadi system of Zelega-Biyar-Ain, as well as Wadi Sa'al, shows a dominant west-to-east flow that coincides with the general slope; however, in the northeast the main course of Wadi Watir is forced into a north-south structural alignment along the great Atiya Fault (Figure 3-7).

The rugged, broken terrain of the Dividing Valleys continues to pose problems for modern transportation. Present highways skirt the edges of the Dividing Valleys, mostly following the valleys of Wadi Feiran and Wadi Watir.

Starting from the northwest, Naqb Rakna provides a rather open access to the Tih Plateau from Wadi Garf. This route was often used by early visitors traveling from Suez to Nakhl and on to Gebel Musa, since the trail is suitable for burden camels.

Three other naqbs, Biyar (also known as "Naqb el Mirad"—Pass of the Watering Place), Thamila, and Suwana, provide local access to the main trail as well as access for nomadic pastoralists between the plateau and the broken grazing lands below (Figure 3-6). Eastward movement beyond the Wadi Akhdar Pass is extremely difficult with laden camels and impossible with motor vehicles.

Plain of Qa

In comparison with the foregoing section the Plain of Qa has a relatively simple morphology. "Qa" literally means "plain," so in effect it is "Plain of the Plain," which is not a bad way to conceptualize this area. From Wadi Hadahid in the north to the neck of Ras Mohammed in the south, it stretches a distance of 130 km. Its maximum width is 23 km just south of the village of Tor. Along the east side of the plain is a narrow pediment, lying against the massif. The wadis which originate in the massif cut across the pediment and mostly die on the plain. In the northwest the plain is separated from the Gulf of Suez by a narrow coastal range.

The surface of the northern section is dominated by a gravel cover. It is windswept and almost devoid of vegetation or wadi channels except in the pediment areas. Just south of Wadi Hebran, where the pediment meets the plain, an outlier of Cretaceous limestone called "Gebel Safarial" is of considerable paleontological importance. It is the only "bone bed" in Sinai.

South of Tor the surface is rippled sand with boulders. Adjacent to the sea there is a narrow but persistent exposure of salty clay. Off shore, coral reefs can be found from Sheikh Raiya to Ras Mohammed.

South of Wadi Letihi, on the east side of the Qa Plain, the

Figure 3-8. The narrow Aqaba Foreshore with the rugged peaks of the southern Sinai Massif in the background. This is Wadi Um Adawi near Ras Nusrani.

sediment-covered pediment gives way to low igneous hills, outliers of the massif. The plain terminates at the neck of the Ras Mohammed headland.

Aqaba Foreshore

The Aqaba side of south Sinai is tectonically more active than the Suez side. As a result the coastal plain from Sharm el Sheikh to Ras Atantur is narrower and more irregular than the Qa. From Ras Atantur to Taba the mountains extend to the water, with only the wadi outlets providing the rudiments of plain development (Figure 3-8). At Dahab the vast wadi system of Nasb and Dahab has built a deltaic plain outward into the gulf. From the air this alluvial shelf stands in bright contrast to the dark hills which otherwise border the crystal blue waters of Aqaba.

Coral reefs on the Aqaba side are more extensive than in the Gulf of Suez, extending beyond the Dahab sector to Geziret Faraun with its fringing reef near Taba. In the area of Sharm el Sheikh elevated reefs of Miocene coral dominate the coastal plain.

Dikes are prominent features in the geomorphology of south and central Sinai (Figure 3-9), but their dominance is nowhere more noticeable than in areas of the Aqaba Foreshore. Extreme denuda-

Figure 3-9. Intrusive gabbro dikes exposed by long periods of uplift and erosion.

tion and the prevalence of felsite and dolerite intrusions prompted W. F. Hume (1906) to label this "Dike Country." He also noted that the dolerite dikes caused his magnetic compass readings to vary by more than 10 degrees. The dikes here are not the longest in Sinai, but they probably display the greatest density of intrusion.

Sinai Massif

To most of us it is the massif of Sinai which holds the greater part of the peninsula's intrigue. History and mythology contribute to a feeling of awe and timelessness, but the great brooding hulks of the mountains themselves, the gebels of the Arabs, are equally capable of attraction with their mystery and foreboding. From the moving vantage point of auto and bus the modern traveler is quickly able to obtain a peripheral and cursory view, but the real personality is reserved for those who enter on foot or by camel.

From the Plain of Qa one can glimpse most of the major peaks of the western range—Serbal, Tarbush, Qasr Abbas Pasha, Zebir (St. Catherine), and Um Shomer. Gebel Musa is hidden by Qasr Abbas Pasha and Zebir. Farther south, as the peninsula narrows and mountain massiveness declines, gebels Thebt and Sahbagh of the dividing range are visible.

Between Wadi Feiran and Ras Mohammed, no highway pene-

Table 3-1. *Distribution of Sinai's Principal Peaks (elevation in m)*

Peaks West of Drainage Divide		Peaks on Drainage Divide		Peaks East of Drainage Divide	
Serbal	2,060	Suweir	1,706	Habashi	1,550
Tarbush (Moreia)	2,093	Um Alawi	2,141	Um Siyala	1,600
Qasr Abbas Pasha	2,383	Abu Mesud	2,135	Feirani	1,685
Musa	2,285	Sabbagh	2,266	Adakkar	1,520
Zebir	2,642	Thebt	2,439		
Um Shomer	2,586	Sahara	1,459		
Rimhan	2,412				

Note: The highest points north of the massif are in the Egma and Tih escarpments. These are Ras el Gineina (1,626), Gebel Dalal (1,606), and Gebel Kheiyala (1,323).

trates the mountain fastness, and only two camel routes connect the Qa with the Gulf of Aqaba. Both routes trace circuitous, southwest-to-northeast patterns through a maze of wadis and passes.

Early travelers to south Sinai frequently approached the central massif via the Wilderness of Tih, descending into the Dividing Valleys through passes such as Naqb Biyar, then up through Watia Pass to Gebel Musa. The modern traveler, however, approaches from the west, following the progression of Wadi Feiran and the foreboding Gebel Serbal, to intersect the ancient Tih route at Watia Pass. From the east a good highway gives access to St. Catherine and the central massif through Wadi Nasb Pass. Farther north, another highway accesses the massif from Nuweiba.

The elevational configuration of the Sinai Massif is asymmetrical, in that seven of its ten highest peaks are found west of the central divide and the other three are on the divide. There are at least seventeen peaks on or west of the divide higher than Gebel Feirani (1,685 m), the highest peak of the eastern portion of the massif (Table 3-1).

Gebel Serbal Area

Gebel Serbal is perhaps the most dramatic peak in Sinai, even though it is thirteenth in elevation. Its rugged profile marks the northwest corner of the massif, directly overlooking the Feiran Oasis (Figure 3-10). From Wadi Feiran the 4-km-long summit appears

Figure 3-10. Gebel Serbal from Wadi Feiran Oasis.

to be dominated by five turret-shaped peaks aligned along an east-west ridge. In reality there are over ten peaks along this summit. The highest, Gebel Madhawwa—the "Lighthouse"—named from the stone structure at its summit, reaches an elevation of 2,052 m before dipping away 1,460 m into Wadi Aleyat (Figure 3-11). Serbal is a typical high, "diked" granite range minutely dissected by rugged, boulder-strewn canyons. To some believers, this pile of granite is thought to be the mountain where Moses received The Law; the consensus, however, favors Gebel Musa as the "true" Mount Sinai.

From Serbal's summit it is possible to see beyond the chaos of granite boulders and expanses of sand to the coastal town of Tor, then across the blue waters of Suez to the purple-red mountains of the African Ranges. Northward is the long line of the Tih and Gebel escarpments. Southeast is a jumble of dark granite mountains laced with thread-like wadi beds winding helter-skelter through the whole. Prominent among these mountains are gebels Zebir and Um Shomer. Gebel Musa lies farther east, mostly hidden among a morass of similar monoliths. Most striking, however, in this desert land is the vista of the oasis at Wadi Feiran, which cuts across the barren waste like a green knife of palms and tarfa, sayal and sidr. No other peak in Sinai lies so close to such a sizeable pocket of biological luxuriance.

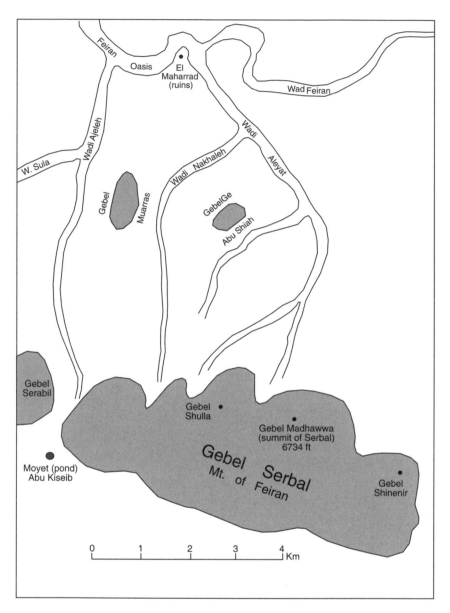

Figure 3-11. The Feiran-Serbal area (adapted from Bartlett).

Gebel Zebir

With the redundance of mountain mythology, Zebir, like Serbal and Musa, is also interpreted as the "Mountain where God Spoke to Moses." Zebir is a group of three peaks sometimes called the "Saint Catherine Cluster." This mountain block is made up of dark syenitic granite, shot through with numerous dikes. Zebir is the central and highest peak at 2,642 m. St. Catherine, slightly lower, sits 1 km to the north. Abu Rumail lies almost 2 km south of Zebir.

From Zebir it is possible to see all the high peaks of Sinai. In the crystal air Musa seems a stone's throw away to the east. To the north lies Qasr Abbas Pasha, and in the distance a little to the left of Qasr, rugged Serbal shows up as a point of light on the distant horizon. Then far beyond, three-fourths of the way up the peninsula, can be seen (on a clear day) the form of Gebel Hamman Faraun. The waters of both gulfs and the escarpments of Tih and Egma are also visible. The high points of this panorama are the great luminous peaks appearing as mountains of light.

Southward the summit mass of Zebir slopes downward into an irregular upland plateau with smaller, shattered peaks of dark syenite that appear almost black. At the distant edge is seen the form of Gebel Um Iswed, "Mother of Blackness." At sunset the effects of color become extravagant. The mix of red-brown, purple, and black from the ancient crystalline rocks contrasts sharply with the blues and lilacs of the sky.

The trail to Catherine follows the Wadi Leja (Figure 3-12) past landmarks with apocryphal names—Mould of the Calf, Burial place of the Tables of the Law, and Stone of Moses (Barron p. 73). The peak itself is a barren block of granite, but its surface is greatly weakened and broken by cliffs and ledges, which makes ascent feasible from almost any side.

Gebel Musa Area

Gebel Musa (Mountain of Moses) is the eighth-highest peak complex in Sinai, but it is the site most widely accepted as the Mount Sinai of the Bible. Musa and its near neighbors, Deir and Fera, are made up of mostly pink to red syenitic granite.

The apex of Gebel Musa sits at its southern extremity. A shallow summit basin separates the high point from Ras Sufsafa, the massive bluffs that comprise "the brow" of the mountain. The difference in elevation is minimal, so that any view of Musa's summit from the Rahah Plain is completely blocked (Figure 3-12).

Below the Brow of Musa and extending northwest past the clus-

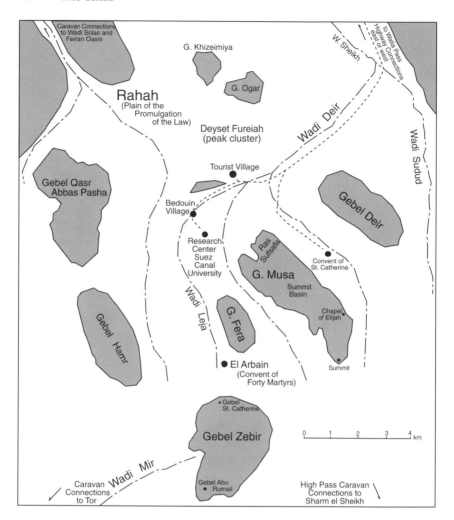

Figure 3-12. The Gebel Musa area (adapted from Bartlett, Hume, and the Tor Topographic Sheet).

ter of peaks known as "Deyset Fureiah" is the Plain of Rahah, the so-called "Valley of the Promulgation of the Law." The plain is some 2 km wide and 3 to 4 km long, sloping gradually upward from a low spot directly below Ras Sufsafa.

Early explorers and Old Testament scholars have attempted to measure the extent of Rahah to determine if it was large enough to hold the Host of Israel to which Moses delivered the law. Estimates were that Rahah proper is 400 acres. Add to this the flat surfaces and lower slopes in Wadi Shreich, Seil Leja, and Wadi Deir, and some 1,200 acres would have been available for the encampment, but only 207 acres would have been needed for assemblage of the Host in the area of the Rahah itself (Barron p. 68).

Throughout this region are numerous permanent springs and small streams. The water supply here is among the most reliable of any area in Sinai. Little wonder that the Host of Israel might have camped here, to say nothing of the host of pilgrims and hermits who followed.

Three main routes now lead to the summit of Gebel Musa. The shortest and most direct trail, the Sikket Syedna Musa (Path of Our Lord Moses), leads southward from the Convent of Saint Catherine through a rugged chute to Elijah's Chapel, then upward over the precipitous granite to the summit (Figure 3-13). This route, traditionally followed by monks and pilgrims, has been fitted with thousands of stone stairs and two nicely fitted stone archways. Constant repair is necessary to compensate for the damage done by rushing water and rock falls. Not far from Elijah's Chapel is the Spring of Moses, where tradition says Moses watered the flocks of Jethro. A second trail from the Gebel Saint Catherine side, called the Pasha's Road, follows a much longer but gentler ascent. The project was an ambitious one. Deep dugways through solid granite still scar the southeast face of the mountain. Though the road failed to reach the summit, it gives camel access to within a few

Figure 3-13. Summit basin of Gebel Musa (Mount Sinai) from the south summit. Gebel Deir can be seen in the distance to the right.

hundred meters, joining the Path of Moses near Elijah's Chapel. A third trail leads from the Suez Canal University Research Center up Wadi Shreich to the south end of the mountain, where it also joins the upper part of the Pasha's Road and finally the Path of Moses for the final ascent.

Six km west and a little north of Gebel Musa is a peak once called "Samr el Tinia," but now known as Qasr Abbas Pasha (Palace of Abbas Pasha). During his brief reign (1849–1854), Abbas, grandson of Mohammed Ali, decided to build a summer palace at the summit of Gebel Musa. The Pasha Road, designed to give access to the summit, was nearly complete when he changed his mind, shifting his efforts to the slightly higher Samr el Tinia. So many Bedouin conscripts lost their lives in the construction of these roads and a sumptuous palace that a legacy of hate continues to this day. The summer following the death of Abbas was so hot that the natives thought the Devil had been extra diligent in stoking the fires of hell to receive their khedive. Some of the extra heat escaped to plague the Sinai.

Gebel Um Shomer

Prior to 1907 Um Shomer was thought to be the highest peak in Sinai. Only the spirit leveling (vertical triangulation to determine the elevation of inaccessible mountain peaks) and barometric observations by Thomas Barron were able to reverse the long-held opinion and place it second to Gebel Zebir. In fact, both Gebel Zebir and its sister peak, Gebel Catherine, are higher than the summit of Um Shomer (2,576 m). The summit is composed of granite so white it is frequently mistaken for chalk, and the lower parts of the mountain are composed of red granite. It is so rugged that J. L. Burckhardt in 1816 was unable to scale the pinnacle. The first ascent was made by three Englishmen in 1862. The view from Um Shomer is much like that from Zebir, except to the south one can look down on Ras Mohammed and the confluence of the gulfs of Aqaba and Suez in the Red Sea proper.

Drainage

Customarily, Sinai is divided into five drainage units: (1) Northern Plains, (2) El Arish Basin, (3) Dead Sea drainage, (4) Gulf of Suez, and (5) Gulf of Aqaba (Figure 3-14). In both area and water yield, the drainage of Wadi el Arish overshadows all other units. The Northern Plains has no drainage channels that reach the sea. The Dead

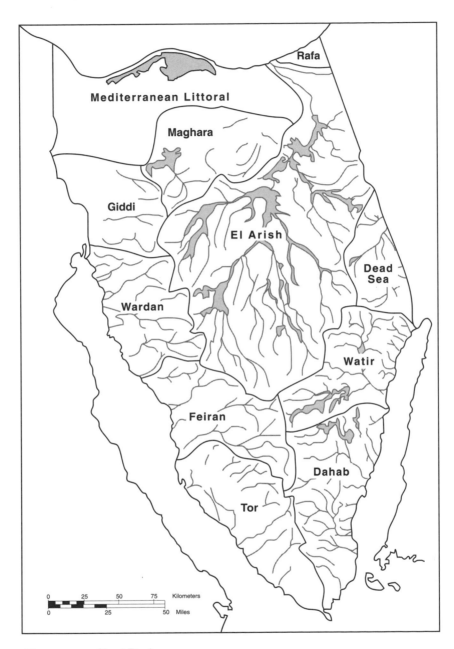

Figure 3-14. Sinai Drainage.

Sea drainage is unique in that its ultimate base level is the lowest elevation on the earth's land surface, but it is outside Sinai proper. As with most desert environments, water flow in the ephemeral streams is at best sporadic. Floods, though infrequent, can be destructive, even while moisture retention is minimal. Water, as might be expected, is the most critical resource and one which is extremely difficult and expensive to husband in a meaningful way.

Wadi el Arish Drainage Basin

Some 28 percent of Sinai is drained by the Wadi el Arish Basin. This wadi is the only stream channel to maintain a course through the great northern dune sheet, and even here no water flows into the sea in most years.

The upland tributaries of the Wadi el Arish essentially all have their beginning along the gentle backslopes leading away from the Tih and Egma escarpments of south central Sinai. These escarpments reach a maximum elevation of 1,600 m, breaking precipitously along the south, but sloping gently downward toward the north. The surface drainage is such that the majority of the precipitation runs into Wadi el Arish rather than southward into the rugged Dividing Valleys drained by wadis such as Akhdar and Zelega. Northward the tributaries cross the plateau, descending as much as 1,200 m (from 1,600 down to 400 m). Three main systems, wadis Bruk, El Arish, and Aqaba (not to be confused with the Gulf of Aqaba), draining thousands of small wadis, converge to form the complex braided channels in broad sand valleys. These channels tend to be broad, sluggish, and sand-choked. There are essentially no tributaries between the Halal Narrows and the Mediterranean. Seldom does water flow in the lower reaches of this important stream channel, but when it does it can be impressive.

Northern Plains

The singular aspect of this drainage unit is the lack of channels. No wadis here reach the sea. The few channels that become established on the flanks of Maghara or Yelleq quickly lose their storm water in the shifting sands of the Dune Sheet. Some water is recovered at scattered springs and wells, but by and large it is devoid of surface water. East of Wadi el Arish the section labeled "Rafah" is really an extension of the northern plains with no permanent channels to the sea.

Gulf of Suez Drainage

Between Uyun Musa and Hamman Faraun only three wadi systems, Lakata, Sudr, and Wardan, maintain channels to the Gulf of Suez. Even the large Gharandal system does not maintain a permanent channel, but loses its flow in the sands north of Gebel Hamman Faraun.

Southward Wadi Garf and Wadi Sidri have permanent channels but seldom run water. Wadi Humr and the noted Feiran system have no outlets to the sea. Their waters are consumptively used or lost into the sands. The significance of Wadi Feiran is its great oasis—by any measure the most productive area in the south. Here Moses fought the Amalekites, we might assume for control of the water and date palm groves, even though they were supposedly living on manna. Its waters are used in a narrow valley often less than 100 m across. The total length is about 7 km. Groundwater reliability is high, so in spite of its restricted size it continues to be one of the most important economic factors in the lives of some southern Bedouin tribes. Unfortunately, as in the rest of Sinai, hydrologic and stream flow data are nonexistent.

Farther south the Qa sand plain is a veritable sink for upland drainage along the west side of the peninsula. The north is fed by seven wadi systems, including Hebran and Mir. The surface flow of all these is lost in the deep sand, but a portion reappears where limestones of the coastal ranges intersect the alluvium. Surfacing waters give rise to the hot springs north of Tor and some wadi flow in the vicinity of Gebel Araba. The hot springs continue to supply spas and some irrigation, but there is no evidence of permanent flow in any of the wadis near the coast.

South of Tor, five main wadi systems feed out of the massif across the piedmont and onto the plain, including Imlaha, Isla, and Thiman. These wadis maintain channels into the gulf even though no water flows in their lower courses in most years.

Aqaba Drainage

The Gulf of Aqaba drainage is logically divided between the Dahab sector, which drains the crystalline massif along its eastern flank, and Watir, which follows a long structural valley. Three main wadi systems dominate the Dahab sector. From south to north these are Um Adawi, Kid, and Nasb. Adawi and Kid have nominal coastal plain development. Farther north, Wadi Nasb (the Nasib-Dahab system) is naturally channelized to the very edge of the tectonic boundary of Aqaba, but its sediment load is such that it has built a deltaic extension over the western edge of the Aqaba pull-apart basin.

North of Dahab, the Wadi Watir dominates the basin and opens onto the gulf at Nuweiba. This basin is fed by the steep south side of the Egma Escarpment and the northeast slopes of the Sinai Massif. Canyons here are extremely rugged, but the presence of reliable springs and intermittent streams helps them accommodate a rather permanent population.

Dead Sea Drainage

North and west of Taba, Wadi Paran forms a section of interior drainage which flows northeast past Kuntilla. This system drains out of Sinai into the Dead Sea (330 m below mean sea level) in Israel. The lowest elevation in the Sinai portion of Wadi Paran is still 480 m above the sea.

Conclusion

By now the reader is well aware of the contrast between the rich geomorphology and the rather severe lack of surface water in Sinai. The constantly moving Dune Sheet of the north, the ancient and stable gravel (reg) surface of the central plateau, and the rugged granite mountains of the southern massif only hint at the diversity, but none of these adds much to the water supply. Numerous dry wadi channels usually carry their infrequent load to the insatiable sands or occasionally into the sea. Only a few favored areas retain enough water to form springs or short intermittent streams, but their occurrence prompts a cultural response that is comparable to the biological richness of the oases so formed. Water is not only limited by factors in the environment, it thoroughly emphasizes them.

4. Weather and Climate

The patterns of weather and climate in Sinai are largely dictated by its position between the vast landmasses of Africa and Asia. Neither the Mediterranean nor the Red Sea significantly ameliorates the extremes of continentality inherited from its large land neighbors. In actuality the continental effect is intensified by the usual east-west orientation of the atmospheric pressure systems extending across north Africa into southwest Asia, which brings the destructive force of great aridity. In the following pages not only will the interaction of climatic elements—insolation, temperature, pressure, moisture, and winds—be considered, but more importantly how they combine to generate the aridity which so dominates this environment.

Incoming Solar Radiation

Solar radiation is the primary element of weather and climate. In varying degrees it affects all other controls. However, its initial influence is upon atmospheric temperature, which in turn modifies pressure and moisture availability. In a moisture-deficient environment like Sinai, the receipt of radiant energy is the most direct control determining the levels of potential evapotranspiration. High potential evapotranspiration and low moisture availability dominate the pattern. Temperature and winds are secondary to radiation as causes of aridity.

Radiant energy receipts for Sinai are high. The annual radiation for the entire region averages more than 500 cal/cm²/day (Table 4-1). A high percentage of Sinai's daylight hours are cloud free, 71 percent for January and 81 percent for July, for an average of 76 percent for the year in the central plateau region (Table 4-2). At El Arish on the Mediterranean coast, cloud-free time averages 74 percent. Sharm el Sheikh on the tip of the peninsula on the Red Sea has 82 percent.

Table 4-1. *Mean Daily Radiation by Months in Central Sinai (cal/cm²/day)*

Jan.	300	July	700
Feb.	400	Aug.	600
Mar.	500	Sep.	550
Apr.	575	Oct.	450
May	650	Nov.	350
June	700	Dec.	300
		Annual mean	506

Source: Thompson (1965)

At St. Catherine Monastery, in the massif with its higher elevation and increased condensation, the average cloud-free time is 70 percent. The most cloud-free area for the whole region is along the Gulf of Suez Coastal Plain, where Tor and Abu Rudeis are both 85 percent cloud free. The effects of the latter will be noticed in high temperature and aridity patterns discussed later.

Atmospheric Temperature

The relationship between solar radiation and atmospheric temperature is obvious. Less obvious is the role of vegetation sparsity and the lack of ground cover in the intensification of atmospheric heat transfer, both heating and cooling. Small amounts of shortwave, solar radiation are converted to longwave heat energy in the atmosphere, but the greatest part is converted at the earth's surface. Thus the solid earth acts as the major conversion system and largely controls the levels of heat transferred to the ambient atmosphere above. In humid climates with dense vegetative cover much of the available heat energy is used in the evaporation of water. Evaporation absorbs heat and provides the potential for greater cloud cover. These processes help to ameliorate the temperature, making it cooler by day and warmer by night.

In dry environments such as Sinai the lack of moisture and cloud cover allows maximum energy conversion at the surface during the day, which is then transferred to the atmosphere by conduction, increasing the sensible heat. Cloudless night skies also allow rapid transfer of energy, but the effect of nocturnal radiation is one of rapid cooling of the lower atmosphere once the sun sets. This latter facet is more significant in winter than summer. Sinai can be extremely hot in summer and very cold in winter, especially in the higher elevations of the southern massif. Maximum screen (shade)

Table 4-2. *Receipt of Sunlight and Percent of Time*
without Cloud Cover by Months for Port Said (31° 17′ N)
and Aqaba (29° 33′ N)

Port Said

J	F	M	A	M	J	J	A	S	O	N	D	YR
226	215	257	279	316	351	357	350	312	295	240	201	3,399 (hours)
71	69	69	72	74	83	82	85	84	83	76	64	76%

Aqaba

J	F	M	A	M	J	J	A	S	O	N	D	YR
230	236	251	224	320	325	348	347	291	282	245	217	3,316
71	75	68	71	67	77	81	85	78	79	76	68	76%

Source: Rudloff (1981)

temperatures for July average 32.5°C, with minimums at 22.5°. In January the maximum and minimum temperatures are 20° and 10°, respectively. World maps of sea-level temperature patterns show a general east-west alignment of isohyets across Sinai. In reality elevational differences modify the expected surface pattern to the extent that isohyets run predominantly north and south (Figures 4-1 and 4-2).

Atmospheric Pressure

On balance the atmospheric pressure over Sinai tends toward the high side, which relates to its latitudinal position, centered as it is across the 30° North Parallel. Air rising from the heat-induced Equatorial Trough of Low Pressure, upon reaching the top of the troposphere, splits and spreads into both hemispheres. This stream of air rapidly cools as it moves along at high elevations just below the tropopause. At approximately 30° north and south the cool air settles downward, producing a high pressure system at the surface. This dynamically induced system is stronger in the winter hemisphere and tends to be more prominent over ocean surfaces than continental masses. The distribution of high pressure affecting Sinai appears as a ridge extending eastward from the semipermanent Atlantic Subtropical High Cell toward the thermally (cold) induced Siberian High Pressure System. Thus the winter high pressure over Sinai is a product of both dynamic and thermal processes (Figure 4-3).

In general, January high pressure over Sinai averages between 1,018 and 1,022 mb (millibars). The prevailing winds are from the

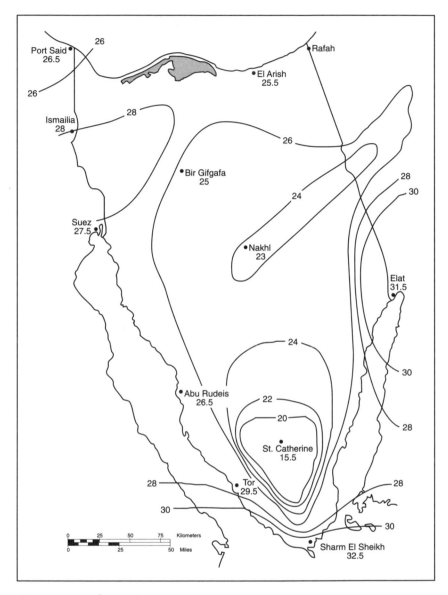

Figure 4-1. July mean temperature.

north, moving from the Mediterranean Sea toward the North Equatorial Trough of Low Pressure, which is centered over the Guinea Coast and Cameroon.

The warming trends of spring cause a shift in the pressure systems. By April the high pressure belt has weakened, so that Sinai generally sits between the 1,010 and 1,014 mb lines. The air flow is

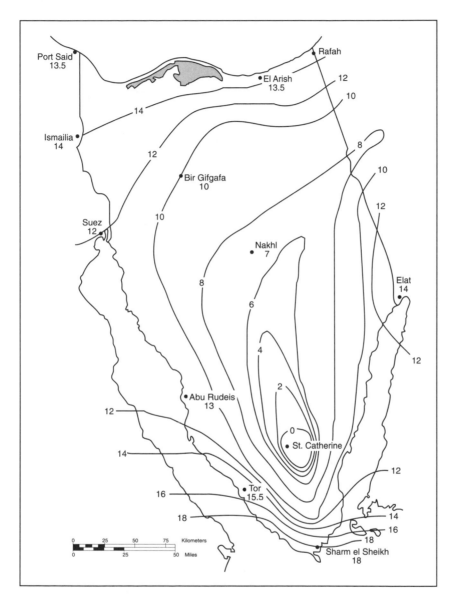

Figure 4-2. January mean temperature.

still predominantly from the north, feeding into an Equatorial Low now centered farther east over Chad and Sudan.

The summer heat of July and August pulls the Equatorial Trough northward, cutting off a cell of moderately low pressure over the western Sahara (Algeria and Mali) and producing a slightly deeper low to the east. Of greatest significance to Sinai in this season is

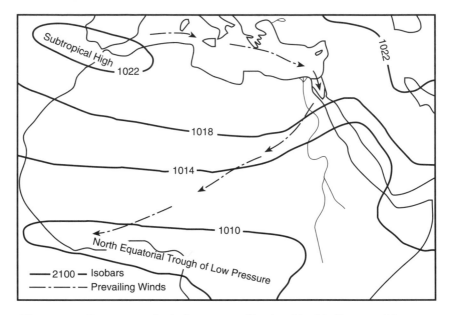

Figure 4-3. Pressure and wind patterns affecting Sinai in January. Note how Sinai sits along a high pressure ridge with higher pressures to both east and west. Source: Thompson 1975, p. 13.

the increased effect of continentality. The vast landmass of Eurasia is now heating steadily. Rising air heated by conduction from the land surface generates a large low pressure region centered over Pakistan and the northern Arabian Sea where pressures dip into the 990's. For Sinai the result is a north-south orientation of isobars, with the entire region lying between 1,004 and 1,008 mb. Prevailing winds continue to blow from the north, but less forcefully than those of spring (Figure 4-4).

Winds

As seen in Figures 4-3 and 4-4, the prevailing winds of Sinai are from the north; however, mountain barriers and deep wadi channels often modify these winds locally. The general air flow from the Mediterranean is the source of nearly all precipitation, the bulk of which falls in winter. With the onset of spring there is an increased chance for strong winds, still predominantly from the north. But the south winds are the most dreaded. These tend to be abnormally hot and dry and accompanied by hazy atmosphere. Such a wind, known as the "Hamasin" for the fifty days in which it is most apt

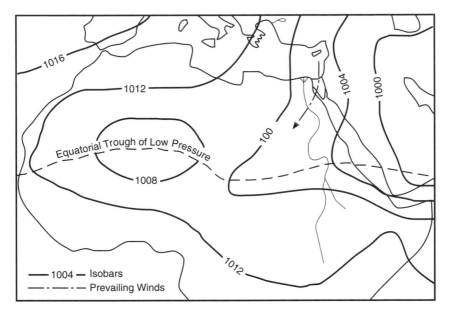

Figure 4-4. Pressure and wind patterns affecting Sinai in July. During the extreme heat of summer Sinai is dominated by low pressure, lying between the 1,004 and 1,008 isobars. Note how the isobars of low pressure are pulled northward by the heating of the vast Eurasian landmass.

to blow, is oppressive and frequently damages vegetation. From year to year the average number of days feeling the full force of these winds are about four. Severe sand and dust storms often accompany the Hamasin.

The end of a Hamasin blow is usually marked by a Shamal, a northerly or northwesterly wind accompanied by a cold front squall, with occasional gales and dust storms. Unfortunately, even when precipitation accompanies a Shamal, it is apt to be of little importance because of increased desiccation and evapotranspiration.

Precipitation

Sinai is a land of little rain. With the exceptions of higher mountains and the tiny northeast corner around Rafah, all areas receive less than 100 mm (3.9 in) of precipitation annually. Over half the landmass, including most of the Tih Plateau, receives only 25 to 50 mm (Figure 4-5). The driest part of the peninsula lies along the Plain of Qa. Tor receives a scant 13 mm (0.5 in) per year. The second-driest area extends up the coastal plain along the Gulf of Aqaba, with

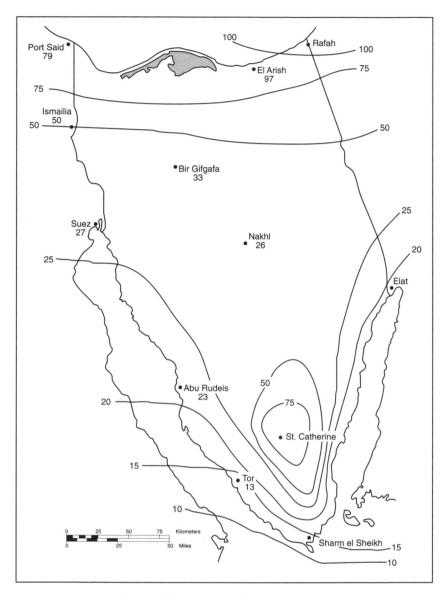

Figure 4-5. Distribution of mean annual precipitation in mm.

receipts gradually increasing from a low of approximately 15 mm at Sharm el Sheikh on the Red Sea to 20 mm at Taba near the head of the gulf.

Along the Mediterranean coast and in the mountains of the southern massif, precipitation is a little higher. At El Arish the total is 97 mm, increasing eastward to 120 mm at Rafah. This reflects the importance of the Mediterranean as a source of winter rainfall. Essentially no summer precipitation falls in Sinai. A southerly flow in winter carries moist air over the coastal plain. At El Arish 87 percent of total precipitation falls between November and March. Four months, June through September, average zero precipitation. Here the mean number of days receiving precipitation in a year is twenty, compared to fifteen at Port Said and five at Nakhl and Taba. It is generally assumed that higher mountains including Zebir, Musa, and Serbal receive more than 100 mm; unfortunately no records are available to verify this. Temperature data are available for the St. Catherine Monastery, but precipitation data are lacking.

Precipitation in Sinai is largely associated with individual storms. Cyclonic movement in its usual form is not well developed. Likewise convective activity and thunderstorms that affect some desert areas are largely missing, since not enough moist, unstable air reaches the area to trigger many storms. In all but the most elevated areas winter precipitation falls as rain; nevertheless snowfall at Musa, Zebir, and other mountain areas can be expected between November and April. Snow accumulation in highly protected, north-facing valleys may occasionally be up to a meter in depth, and some snow cover can be expected through January and February.

As in most desert environments, rainfall variability is high, meaning reliability is low. All of Sinai has variability exceeding 40 percent. Low and variable precipitation by themselves do not provide a particularly accurate measure of the real problem. Only when precipitation is related to the potential for evapotranspiration do we find an accurate measure of aridity.

Aridity

The Bedouin insist that the poor feel the cold more than the heat, but concede the sun to be the greater enemy that attempts to destroy them by drying up all the moisture. For the geographer who does not shiver under the goat hair tent in January, the tendency will be to relate to the second threat as the dominant factor of weather and climate in Sinai.

Modern climatic theory depends upon a number of multi-variable phenomena such as the relationship of precipitation to potential evapotranspiration (aridity) or that of temperature and air movement (wind chill) to enhance the accuracy of our perceptions. Indeed, to modern climatology the balance of precipitation and potential evapotranspiration determines the boundary between humid and arid climates, as well as the intensity of either. Aridity, in spite of the Bedouin dread of cold, must here be the central focal point. In 1977 UNESCO (United Nations Educational, Scientific and Cultural Organization) conducted a worldwide study of aridity. According to this study all of Sinai falls into two classes: "arid" (precipitation–potential evapotranspiration ratio of .03–.20) and "hyperarid" (ratio less than .03). Thus the arid category includes areas where precipitation is only 3 to 20 percent of the potential evapotranspiration. Hyperarid areas receive levels of precipitation less than 3 percent as great as potential evapotranspiration. For example, in central Sinai at Nakhl the average precipitation is only 26 mm (1 in) per year, but the radiant energy available could evaporate 2,500 mm (98 in), giving a precipitation-to-potential-evapotranspiration ratio of .01, making it hyperarid.

The UNESCO division of Sinai into "arid" and "hyperarid" is illustrated in Figure 4-6. Along the Dune Sheet of the Mediterranean Littoral and in the Sinai Massif the climate is arid. In the north the Mediterranean Sea helps to produce winter precipitation, but in summer the influence of continentality eliminates the moisture and drought is maximized. This is also a "xeric moisture regime," which is defined as one having an absence of growth moisture in the soil for a period of forty-five days or more in the four months following the summer solstice. In the mountains of the south a similar pattern exists, but elevation rather than the ameliorating influence of the sea provides the extra precipitation that keeps it out of the hyperarid category.

In the center and south periphery covering the Tih Plateau and the coastal plains of Suez and the Aqaba Foreshore is found the "hyperarid" zone. The entire area comes under an extremely intense xeric moisture regime, but retains the mild winter temperature patterns found in the north.

Köppen Climatic Classification

Worldwide, the best-known system of climatic classification is that developed by Wladimir Köppen. Köppen was a botanist as well as a climatologist, which was the reason his initial system, published in 1919, used vegetative names such as "rainforest," "tundra," and "desert" to designate various climatic classes. His 1923 comprehensive system substituted letters like "Af," "ET," and "BW" for the original names in an attempt to divorce the empirically derived formulas from a vegetative context. The names, however, in most instances stuck, and the tendency is to use both types of designation.

Within this system all of Sinai is a climatic desert. No humid or

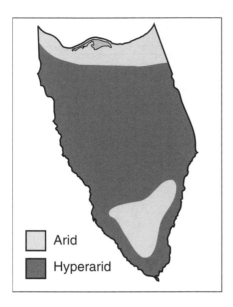

Figure 4-6. Aridity based on the 1977 UNESCO study. Arid—Precipitation–potential evapotranspiration ratio .03 to .20. Hyperarid—Precipitation–potential evapotranspiration ratio less than .03.

Arid

Hyperarid

even semiarid climates exist here, yet there is variety within the extremes of aridity. Based on the Rudloff application of the Köppen system, three broad categories are found: Deserts (BW), Marine Deserts (BM), and Desert Mountains (GBW). It should again be stressed that Sinai is totally desert; not even semiarid steppes (BS) are found in the macroclimatic pattern. Microclimates will be discussed later.

The advantage of the Köppen system is its ability to delimit climatic classes using readily available empirical data of mean monthly temperatures and precipitation. The "dryness limit" (RB) is the boundary between semiarid steppe and humid climates. RB is determined by a simple formula, RB = T-10 + 0.3PS—where T = the mean monthly temperature for the year and PS = the percentage of summer precipitation (April through September). RB is also = to the unity of precipitation and potential evapotranspiration. Any station where precipitation is lower than RB is arid. A second limit, the "desert limit" (RW), is then introduced. This value is one-half of the RB and provides the boundary between the semiarid steppe and the fully arid desert. Thus a semiarid climate would exist where precipitation is between half of potential evapotranspiration and the full value of potential evapotranspiration. Table 4-3 provides comparative data for precipitation and RW (desert limit) values for Sinai stations. In addition it shows a wetness ratio—i.e., the level of precipitation as a percentage of the environment's ability to evaporate moisture if it were available. Consider

Table 4-3. *Climatic Stations of Sinai Showing Desert Limits, Wetness Ratio, and Köppen Climatic Class*

Station	T Mean Annual Monthly Temp. °C	R Total Prec. mm	Summer Prec. mm	PS % Summer Prec. mm	RW Desert Limit mm	Wetness Ratio Prec. % of PET	Köppen Class
Port Said	22	63	6	9.5	149	21	BMhl
El Arish	21	97	8	8.2	135	36	BMal
Ismailia	21	44	7	16	158	14	BWhl
Bir Gifgafa	18	33	5	15	125	13	BWal
Suez	23	27	5	6	147	9	BWhl
Nakhl	17	26	5	19	109	12	BWak
Aqaba (Elat)	25	27	3	11	183	7	BWhl
Abu Rudeis	22	23	3	13	159	7	BWhl
Tor	21	13	0	0	110	6	BWhl

Note: The difference between R and RW is indicative of the intensity of aridity for any station. R would have to exceed RW for a station to be in the semiarid or humid class. Thus Nakhl receives an average of 26 mm/year; it would have to receive in excess of 109 mm to change from a desert to a semiarid climate and 218 mm to become humid.

Source: Rudloff (1981)

that at the boundary between the desert and steppe the wetness ratio would be exactly 50 percent. Note that all of Sinai is drier than the RW limit, so the lesser value of RB is not shown.

BM (Marine Desert) climates differ from traditional deserts (BW) in two ways. They are less dry—note the wetness ratios for Port Said and El Arish on the Mediterranean coast are 21 and 36 percent, respectively, compared to 6 and 7 percent at Tor and Abu Rudeis. Secondly, the relative humidity is considerably higher, with an annual range of 68–72 percent at El Arish, compared to a range of 30–44 percent at Tor.

When the wetness factors have been determined, the stations are evaluated for temperature. A lower case letter is used to designate summer and winter conditions according to the universal thermal scale (Table 4-4).

Unfortunately complete meteorological data are lacking for mountain stations (GBW), but temperature and humidity data for St. Catherine Monastery partially illustrate the nature of these stations. Precipitation below the RW limit means they are true deserts, but they have cool to cold winters. At the monastery, the Feb-

Table 4-4. *The Universal Thermal Scale*

Temperature °C	Description	Köppen Symbol
35 +	severely hot	i
28 to 34	very hot	h
23 to 27	hot	a
18 to 22	warm	b
10 to 17	mild	l
0 to 9	cool	k
−9 to −1	cold	o
−24 to −10	very cold	c
−39 to −25	severely cold	d
—— to −40	excessively cold	e

Source: Rudloff (1981)

ruary temperatures average freezing, and the summers have a mild warm month of 17°C (63°F).

Month	J	F	M	A	M	J	J	A	S	O	N	D	YR
Temp	3	0	2	7	10	13	15	16	17	6	12	7	10

The monastery would be classed GBWkl.

The warmest station, Sharm el Sheikh, has a July mean temperature of 31°C and a January mean of 18°C. Note that the cold month mean is higher than the warm month mean for St. Catherine Monastery. These very high temperatures put Sharm in a slightly different class than Sinai's other hot, sea-level stations—BMhb. The distribution and comparative characteristics of Sinai's climates are shown in Figures 4-7 and 4-8.

Water Budget Climatology

The effects of climate extend below the atmosphere to the edaphosphere (soils) and biosphere. The effectiveness of moisture in a given environment throughout the year is readily quantified in a water budget diagram. The water budget formula is written $P = AE + S \pm ST$—where P = precipitation, AE = actual evapotranspiration, S = surplus, and ST = change in stored soil moisture. For humid or semiarid environments, some portion of the year has a

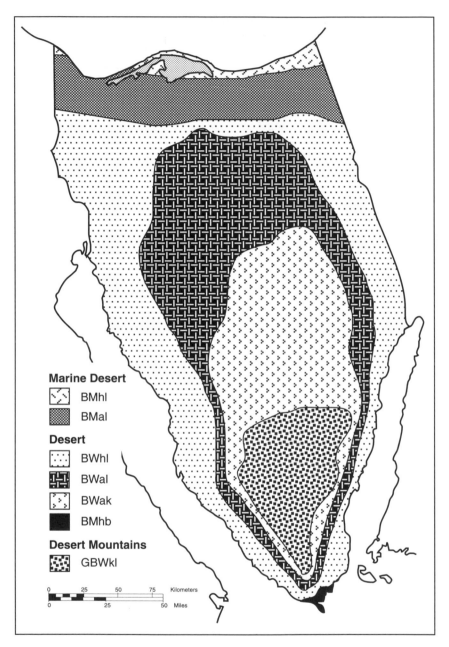

Marine Desert

- BMhl
- BMal

Desert

- BWhl
- BWal
- BWak
- BMhb

Desert Mountains

- GBWkl

0 25 50 75 Kilometers
0 25 50 Miles

Figure 4-7. Sinai's climates based on the Köppen system of classification (Rudloff's modification).

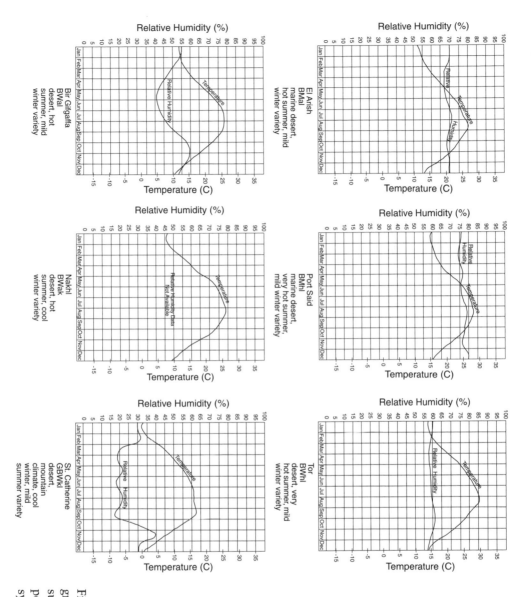

Figure 4-8. Climatic graphs illustrating six subclasses of the Köppen/Rudloff climatic system found in Sinai.

surplus of moisture, i.e., more precipitation than potential evapotranspiration. This surplus provides for maintenance or replenishment of soil moisture reserves and additional stores added to the groundwater supply. In arid climates there is no surplus, so actual evapotranspiration is the same as precipitation. The water budget formula is in effect shortened to P = AE. A soil moisture deficit exists at all times. Thus plants can only produce during the rainy season. No residual of soil moisture exists to support plant growth into the dry season. Figure 4-9 illustrates the water budget characteristics for three arid Sinai stations. As Sinai has only arid climates, a semiarid station from outside is shown for comparative purposes. On the average these arid environments have a minimum of soil moisture replenishment or groundwater recharge during the year. Only the extreme storms carry enough water to add to these stocks, and even then the positive results are apt to be felt at some distance from where the moisture fell.

Importance of Microclimatology

The term "allogenic water" is used by biogeographers to describe water derived from a source outside the area of immediate consideration. Herein is found the most important aspect of microclimatic development in Sinai's deserts. Oases and ains (wells) are dependent upon the transfer of water, usually from adjacent uplands or along protected wadi channels, where deep, coarse alluvium conducts and insulates water left by infrequent storms. Thus movement and storage are dependent upon well-developed and critically placed aquifers where hydrostatic pressure is able to maintain a dependable supply. Closely associated with allogenic water are the protective functions of landforms that hold back the movement of unstable sediments and ameliorate the input of extreme temperatures and direct radiation. In effect, these sheltered microclimates hold the levels of moisture availability in line with ambient potential evapotranspiration, so that even water-loving plants such as date palms are able to thrive and produce.

At the opposite extreme are those microenvironments which support no vegetation whatsoever—desert pavements and persistently shifting sands. There are no significant areas in Sinai that can be classed as absolute biological deserts in a macroclimatic context; rather, it is the microenvironments that fail to provide stable rooting mediums or deny access to water, thereby controlling the growth of even the most hardy xerophytes. Indeed the drought-resistance of some xerophytes is such that only the harshest microclimates caused by physical factors preclude plant growth.

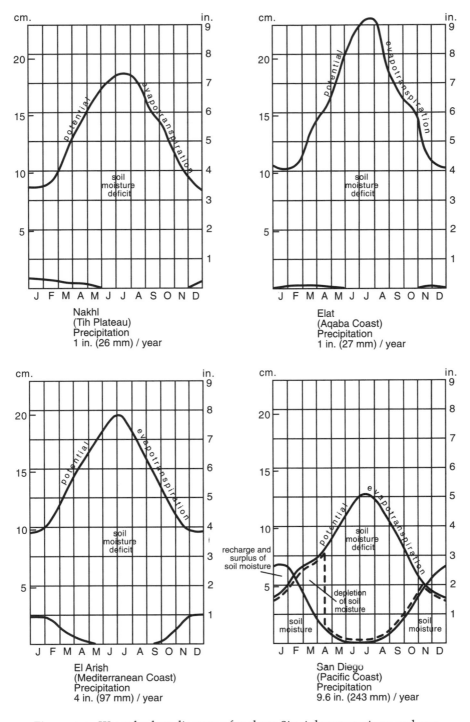

Figure 4-9. Water budget diagrams for three Sinai desert stations and one steppe environment (San Diego, Calif.), given for comparative purposes. The water budget formula—P = AE + S ± ST (precipitation = actual evapotranspiration + surplus ± change in stored soil moisture). Potential evapotranspiration is based on the Blaney-Criddle Formula.

Bedouin Climatology

"The moon condenses the water vapor, distills the dew, and attracts the rain, but the sun attempts to destroy the Bedouin by drying up all the moisture." Among the pastoralists of the Near East, there is a vast lore of weather and climate. This folk knowledge, accumulated by nomadic peoples who depended upon their flocks for sustenance, is not only logical but pertinent to their survival.

The Bedouin year is traditionally divided into five seasons. The year begins with Assferi, an extended autumn. This is the expected rainy season, a period of about ninety days extending from early October to early January. Assferi begins when Canopis becomes visible in the southern sky. Canopis reigns (is visible) for about forty nights in the latitudes of Sinai. It is then that the autumnal rains, "Alwasm," are expected. Of this time it is said, "trust not the wadi and gather the dates even at night" (Musil 1928). The following fifty days, still a part of Assferi, are divided equally between the time of Pleiades and that of Gemini. It is the rains of Gemini that primarily determine the quality of grazing in the coming months. This is the season when the Bedouins should leave their settlements, taking the herds into the desert in anticipation of new grass.

Assferi is usually the cloudiest season of the year, with "mizen" (small clouds) sometimes grading into "seheb," the heavy rain clouds. Only infrequently do the "sana'at," the thunderheads, develop. When they come, however, they produce the heaviest and most destructive rains of all.

Assta, the second season, is a shorter, forty-day period that would be considered winter. Sirius now dominates the sky. In good years the rains of Gemini have already soaked the soils and triggered the growth of grasses, so that the rains of Assta are able to fill the ponds and put the finishing touches on a good season for the herds and herders alike.

From mid-February to mid-April is the fifty-day season called "Assmak." Arcturus is now prominent in the evening sky of early spring. Rains of Assmak are beneficial only if the soil has been thoroughly soaked by the rains of autumn.

Asseif comes quickly on the heels of Assmak, running from mid-April to early June. Westerners would probably think of this as spring, but for the Bedouin it is summer, perhaps better thought of as the "good summer." If rains have fallen, it is a season of abundance and the best time of the year—the time before Al-kez.

Al-kez, "the terrible summer," comes in June and lasts until October. It is the longest and most difficult season of the year, certainly the most dreaded. If the range is good and the ponds full to begin with, it is readily endured, but in years of winter drought it will take an extreme toll on livestock and the well-being of the people.

But things are changing. The Bedouin are caught in a bind of politics and economics. Pressures toward sedentarization and population growth and recent heavy losses of livestock in years of extreme stress have conspired to restrict the wild, free life of the Sinai tribesmen. Toyota trucks and Peugeot sedans compete with the camels, and concrete block houses often replace frond windbreaks and goat hair tents. Bedouin climatology is less and less important to survival and is gradually being lost to Western calendars and commercial work weeks.

Putting Climate to Work

Rapid changes in traditional livelihoods and the trend toward settling Bedouins in permanent villages may create nostalgic hangups, but there are also opportunities. Now is the time for developing a new climatological awareness, one geared to the energy needs of permanent housing, water purification, and even the generation of electricity. Centuries-old problems of overgrazing, charcoal burning, and shivering under goat hair tents should not be solved by increasing reliance upon the fossil fuels which Sinai contains. At least some partial answers must lie in the development of passive solar conversion and other simple noncombustion technologies. For further discussion of this topic see "Housing to Put Climate to Work" in Chapter 7. A new trend in Bedouin climatology should be in the making. If the ancient lore of climate helped herders live with the harsh conditions of their free nomadic life, there is no reason why the development of a new lore should not be useful to a sedentary lifestyle.

5. Soils of Sinai

Detailed soil surveys are one of the crying needs in Sinai's push to develop in appropriate and sustainable patterns. Governor Moneer Shaesh of the North Sinai Governorate was among the first to confirm that the push to bring more fresh water into the peninsula could be highly wasteful without detailed soil information to guide where the water should be used. Many young and well-trained scientists are working to reduce these problems. Soil scientists, engineers, and water resource specialists, with little monetary return but a great love for the land, have chosen to devote their energies to a better utilization of Sinai's fragile resources.

Reconnaissance

The reliability of soil mapping, that is, the accuracy of the maps themselves, is based upon the sources used. In the United States soil surveys for individual counties are generally based upon second- and third-order surveys. Most agricultural land and other areas of high-intensity use are given a second-order survey, which maps soils at the series and phase level. A soil series is the most specific level of classification and would be comparable to the species level in the classification of plants and animals. Such detailed surveys are plotted on large-scale aerial photographs which show roads, buildings, field borders, and even individual trees. This information is then slightly generalized and printed on maps at a scale of 1:24,000. These can be used in conjunction with topographic maps of the same scale. Such detail shows boundaries with considerable accuracy. The phase of a given series is associated with steepness, rockiness, or other topographic difference that may alter the use to which a given soil may be put. Thus modern soil maps at a scale of 1:24,000 are fairly accurate slope maps as well.

Third-order soil mapping in a given area is often used to cover largely barren mountains, extensive dune sheets, and other lands unsuitable for intense development. Base maps are usually at the same scale as those used in second-order surveys. Smaller-scale maps may show soils in much broader categories such as orders or great groups.

FAO/UNESCO's "Soil Map of the World," printed at a scale of 1:5,000,000, shows soils at extremely generalized levels of sub-order or great groups, which is useless in the field. Such data are compiled from a variety of sources, depending upon availability, including: (1) soil surveys (as described above), (2) exploratory soil studies, and (3) spot surveys. In the case of Sinai the map was entirely dependent upon the last category.

This UNESCO map uses nine classes and is the basis for Figure 5-1. However, the terminology used by UNESCO has been modified to the classification system (*Soil Taxonomy*) developed by the Unites States Soil Conservation Service.

The General Map

Figure 5-1 shows the distribution of Sinai's dominant soils. In this worldwide classification there are eleven major groupings, called "orders." Two of these soil orders, Entisols and Aridisols, dominate Sinai. Entisols, those with little profile development, are the more extensive. This is not surprising considering the relationship between climate and geomorphic processes that occurs here. Aridisols, soils reflecting the influence of arid climates and desert vegetation, are more scattered but still significant. Neither order is ideally suited to agriculture, but with proper care and drainage they can be made productive with irrigation.

Entisols

Entisols are chiefly characterized by lack of horizon development. They usually have no surface diagnostic horizons—designated "A." Desert Entisols are little affected by pedogenic processes because they have been so recently deposited or so disturbed that there has been insufficient time for pedogenesis to occur. The Entisol order is divided into five suborders. Three suborders, Psamments (sandy soils), Fluvents (stream deposited soils), and Orthents (true Entisols), are significant in soil development in Sinai. The Entisol categories delineated in Figure 5-1 are discussed in greater detail as follows:

- Dunes (Erg) and Quartzipsamments—Much of Sinai's northern littoral is covered by fine aeolian sands that have no profile development. Quartzipsamments are Entisols (recent

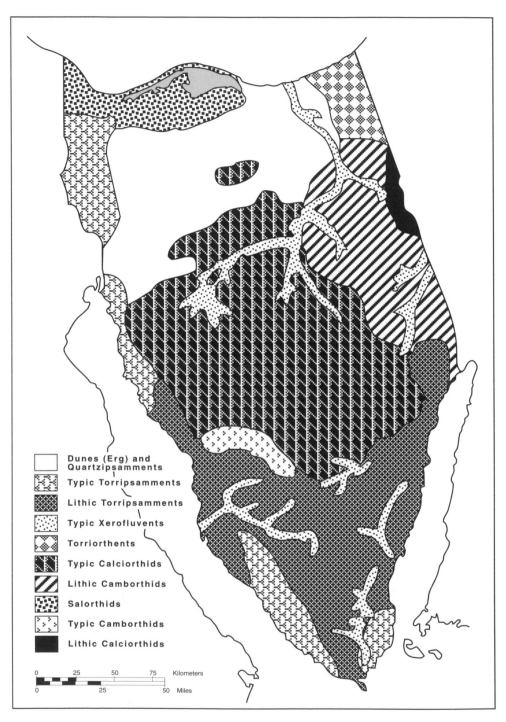

Legend:

- Dunes (Erg) and Quartzipsamments
- Typic Torripsamments
- Lithic Torripsamments
- Typic Xerofluvents
- Torriorthents
- Typic Calciorthids
- Lithic Camborthids
- Salorthids
- Typic Camborthids
- Lithic Calciorthids

0 25 50 75 Kilometers

0 25 50 Miles

Figure 5-1. The principal soil regions of Sinai.

soils, having no pedogenic horizons) with no lithic contact within 25 cm of soil surface, and have slopes < 25 percent and low organic carbon content (typically < 0.2 percent) that decreases irregularly with depth. These soils have a sand fraction that is 95 percent or more quartz and related silicon minerals which do not liberate Fe or Al upon weathering. These soils dominate about 18 percent of Sinai's surface.

- Typic Torripsamments—Sandy Entisols much like Lithic Torripsamments but lacking a lithic contact (bedrock) within 50 cm of the surface. These soils dominate approximately 9 percent of the surface.
- Lithic Torripsamments—The mountainous areas of southern Sinai have much bare rock surface, but soil accumulations in wadis and other sheltered areas are dominated by Lithic Torripsamments. These sandy Entisols have a Torric moisture regime (never moist in some or all parts of the soil for as long as ninety days when the soil temperature at 50 cm is above 8°C). They are shallow, having a lithic contact (bedrock) within 50 cm of the soil surface. These soils dominate in some 17 percent of the surface areas.
- Typic Xerofluvents—Alluvial soils of recent origin (Entisols) deposited and frequently reworked by the large wadis. These soils are dominated by sands and have an Ap (disturbed) surface horizon. They tend to be well drained and nonsaline. They can be moderately productive with irrigation and careful management. Most of Sinai's potentially best agricultural soils are found in this group (Figure 7-2). In total they dominate about 6 percent of the soil surface grouping.
- Torriorthents—Sand soils of the Yammit-Rafah area that tend to have high calcium contents. With proper management these can be adequate agricultural soils. They cover less than 3 percent of Sinai, but they could be important to future irrigation development.

While no Sinai soil is highly fertile, there are some areas of fair to good soils primarily based upon alluvial deposition in the broad wadi systems of the Tih Plateau, the lower drainage areas of the Wadi el Arish, and the Qa Plain. Figure 5-2 locates the most suitable of these soils and shows highway access to them.

Entisol Profiles of the Wadi el Arish

The specific profiles used in this study were taken from Son el Bricki on the main stem of Wadi el Arish some 17 km south of the

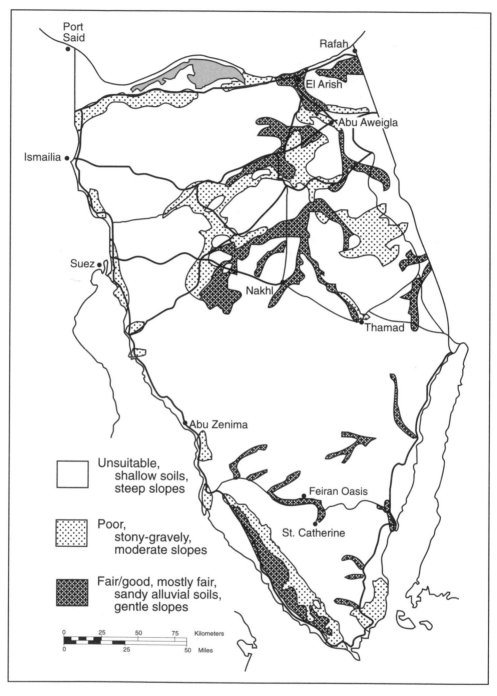

Port
Said

Rafah

El Arish

Abu Aweigla

Ismailia

Suez

Nakhl

Thamad

Abu Zenima

Unsuitable,
shallow soils,
steep slopes

Poor,
stony-gravely,
moderate slopes

Fair/good, mostly fair,
sandy alluvial soils,
gentle slopes

Feiran Oasis

St. Catherine

| 0 | 25 | 50 | 75 | Kilometers |

| 0 | 25 | | 50 | Miles |

Figure 5-2. Cultivable soils of Sinai, showing highway access (adapted from Dames and Moore).

city. In the past this area has had considerable military use, as, for example, bunkers and mine fields, the latter ostensibly marked by barbed wire barriers. Care was necessary to keep student research assistants from accidently excavating in mined areas (Figure 5-3).

Son el Bricki is a typical cross-section on the mainstem of Wadi el Arish, and the soils are representative of Sinai development. Here the floodplain is 1.5 to 2 km wide and somewhat irregular, with relatively flat areas interspersed with dunes of up to 2.5 m in local relief. Some dunes have been stabilized by vegetative growth, with species of tamarisk being the most common agents. Along the east side of the wadi channel is a cut-bank typically 1.75 m high that separates the lower and upper floodplains (Figure 5-4). This bank provided the most complex soil profile found in the area (Figure 5-5, No. 3). Beyond the floodplain to the east is an irregular, undulating terrace largely covered with recently deposited Quartzipsamments. West from the wadi the slope is more abrupt, rising to rather substantial sand mountains also covered with Quartzipsamments, but mostly weathered in place.

Within the floodplain the Bedouins practice a scattered, rain-fed agriculture, growing barley, melons, and other crops that can germinate and mature with the small amount of soil moisture stored from scant winter precipitation (see soil moisture data, Table 5-1).

Figure 5-3. Bazooka rockets with live explosive heads, one more reason for careful research in Sinai.

Figure 5-4. Students from the Faculty of Environmental Agricultural Sciences examine soils at Son el Bricki.

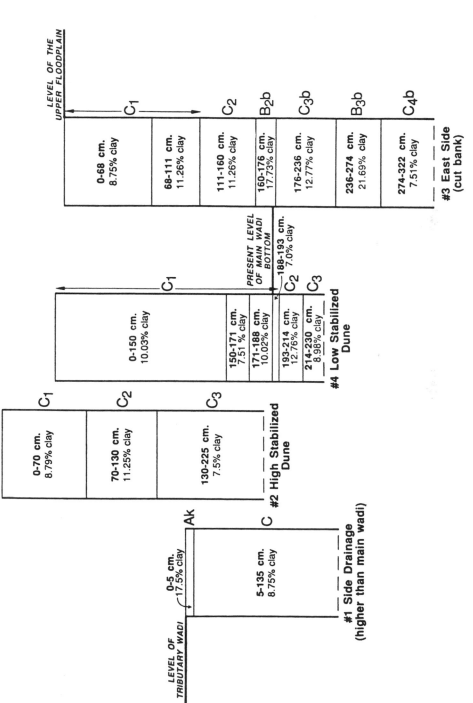

Figure 5-5. Entisol profiles from Son el Bricki, Wadi el Arish.

Table 5-1. Laboratory Analysis and Munsell Color Notations for Entisol Profiles from Son el Bricki

Profile Number	Horizon	Depth (cm)	EC mmohs/cm	Organic Matter %	Ph	CEC Meq/100g	Mechanical Analysis Clay %	Silt %	Sand %	Munsell Color Dry	Moist
1	Ak	0–5	0.54	0.040	7.04	8.2	17.50	1.25	81.25	10 YR 7/4	7.5 YR 6/4
	C	5–135	0.50	0.053	7.24	6.2	8.75	1.25	90.00	10 YR 7/4	7.5 YR 6/4
2	C1	0–7	1.56	0.053	7.97	6.5	8.79	3.77	87.44	10 YR 7/4	7.5 YR 5/4
	C2	7–130	1.20	0.053	7.54	7.5	11.25	2.50	86.25	10 YR 8/4	7.5 YR 6/4
	C3	130–225	1.20	0.080	7.86	5.3	7.50	2.50	90.00	10 YR 7/4	7.5 YR 5/4
3	C1	0–110	0.54	0.007	7.33	6.5	8.75	6.26	84.99	10 YR 8/4	7.5 YR 6/4
	C2	110–160	0.50	0.007	7.25	7.1	11.26	25.03	63.71	10 YR 8/4	7.5 YR 6/4
	B2b	160–176	0.61	0.053	7.18	8.3	17.73	6.33	75.94	10 YR 8/4	7.5 YR 7/4
	C3b	176–236	0.73	0.007	7.47	7.7	12.77	5.11	82.12	10 YR 7/4	10 YR 6/4
	B3b	236–274	0.72	0.027	7.30	9.3	21.69	15.31	63.00	10 YR 8/3	10 YR 6/4
	C4b	274–322	0.78	0.027	7.30	5.6	7.51	1.25	91.24	10 YR 7/4	7.5 YR 6/4
4	C1	0–193	0.90	0.013	7.26	6.9	10.03	1.25	88.72	10 YR 7/4	7.5 YR 5/4
	C2	193–219	0.60	0.080	7.30	7.6	12.76	6.38	80.86	10 YR 7/4	7.5 YR 6/4
	C3	219–230	0.54	0.007	7.30	6.2	8.98	1.28	91.02	10 YR 7/4	7.5 YR 6/4

While we sampled here, a Bedouin farmer with a draft camel and a seeding plow was planting sweet melons in the lowest part of the floodplain, immediately adjacent to the wadi bottom between profiles 3 and 4 (Figure 5-5). The rows were spaced about 1.5 m apart to maximize soil moisture and allow for vine spreading as the crop grew. Higher in the floodplain, east of profile 3, barley had been planted, but had not yet germinated.

Taxonomy

The soils of the wadi margin are typical of the Entisols order in Sinai. The Reconnaissance Map of Sinai Soils (Figure 5-1) shows this area as having Xerofluvents derived from mixed parent materials. The suborder "Fluvents" indicates alluvial deposited materials. The Great Soils Group "Xer" indicates development under a xeric moisture regime in which the soil contains no growth moisture for forty-five days in the four months following the summer solstice. These soils are actually Torrifluvents. The difference between Xerofluvents and Torrifluvents is soil temperature, which affects moisture loss and storage. In xeric regimes the average annual temperature is below 22°C. In torric regimes it is higher, with a concurrent increase in overall aridity. No direct measurements of soil temperatures are available for Son el Bricki or El Arish, but the mean annual temperature of 26°C would almost certainly result in a soil temperature in the lower part of the torric range. For these borderline areas the difference is so minor that it would have little effect on water needs or management practices. In other parts of the peninsula with lesser precipitation or higher temperatures there may be significant differences.

While alluvial soils dominate the floodplain there are also evidences of aeolian (wind) action. Dunes are common throughout, but show greater vertical development along the west side of the wadi. Many dunes have been stabilized by vegetation.

Profiles

As expected, there is a high degree of similarity among the profiles sampled at Son el Bricki (Table 5-1). Textures are consistently sand to sandy loam. Cation exchange capacities are low, ranging from 5.2 to 9.3. Reaction (pH) ranges from 7.04 to 7.97. Only profile 3 shows signs of pedogenic structural development. Colors are typically light, showing strong reflective values. They are mostly pink to reddish-yellow and light reddish-brown. Figure 5-6 shows profile designations. In spite of differences, all four profiles at Son el Bricki properly belong in the subgroup Typic Torrifluvents (Figure 5-1).

Profile 1—was the most shallow and also the simplest one tested. It was located in the bottom of a small side channel adjacent to the main wadi. Its primary distinguishing characteristic was a thin surface layer (5 cm) showing calcium carbonate accumulation to the degree of weak cementation. This has been classified as an Ak horizon. A single C horizon is uniform throughout the remainder of the sampling pit. The drainage for this site was excessive and almost no soil moisture remained. The only surprising aspect of this profile was the low EC (electrical conductivity) of .54 in the Ak horizon. A greater salt content was expected.

Profile 2—comes from one of the higher stabilized dunes. It is a relatively uniform profile with three simple C horizons. The middle horizon had enough fines to class it as loamy sand rather than sand, as in the other two. Organic matter throughout all sample layers of all profiles was extremely low. The C3 horizon of this profile was tied for the highest organic matter content of all samples taken—.08 percent. Other samples ranged downward to a low of .007 percent. It was expected that this profile would be higher in organic matter than others because of the tamarisk stabilizing the dune. In fact, however, there were very few roots and little dead organic matter throughout.

Profile 3—is the most complex profile sampled at Son el Bricki. Significant increases of clay at a depth of 16 to 176 cm and again at 236 to 274 cm levels are strongly indicative of buried alluvial horizons. The slight increase of organic matter in the 16 to 176 cm layer (B$_{2b}$ horizon, the buried remnant of an older soil) bears no relationship to the present sandy layers above. It appears to have been derived from clay translocation (downward movement) from a humic epipedon (A horizon) that has since been truncated. Recent alluvial depositions over the truncated profile have produced the C1 and C2 horizons at 0–68 cm and 68–110 cm, respectively.

Profile 4—like profile 2, was cut through a stabilized dune, but one of lower elevation and immediately adjacent to the main wadi channel. In most ways it was quite comparable to profile 2. There was a similar lack of organic matter throughout the profile in spite of vegetative stabilization.

Dune Stabilization. Prior to exposing the profiles for sampling, it was expected that root structure and organic matter accumulation would play an important role in the retention of soil materials in stabilized dunes. There was, however, a minimum of both live roots and dead organic matter. Without further investigation we can only guess that the modification of wind patterns by the above-ground structure of the plants is the most significant factor in dune stabilization at this site. The dune in turn insulates soil moisture pockets at the plant root zone, thereby forming a complementary system. Thus deflation and deposition by aeolian forces conforming more to the vegetative structure appear to be more important to dune stabilization than the trapping of soil by plant roots to retain the soil against wind erosion.

Agricultural Potential.   The bottom line value of any soil study in Sinai is probably the determination of the potential to produce crops. Sinai Entisols have low to moderate productive potential. They are extremely low in organic matter, which strongly affects water-holding capacity and fertility. All samples showed less than .1 percent organic matter—the actual range was .007 to .08 percent. This problem has been compounded by millennia of overgrazing.

The cation exchange capacity (the fundamental measure of fertility) is also very low, typically only 6–8 meq. This compares with 75–100 meq in good prairie soils. Since the soils are low in clay and high in sand the only logical way to counter this problem is through increased organic matter (humus has a high exchange capacity). Unfortunately the organic matter deficiency has no easy solutions.

On the positive side, these soils are well drained and have good tilth, which makes them easy to work. Frequent deposition of fresh materials by floods helps in a small way to alleviate some of the fertility problem—unfortunately, it is not enough.

Management.   Should these soils be subject to irrigated agriculture, fertilizers will become a major cost factor and care will be required in their organic matter management. The problem is compounded in that leveling and terracing practices must restrict flood sediment renewal. Conversely, insufficient prevention of flooding will subject the soils to erosional losses.

The Storie Index.   Of the various methods of rating agricultural soils, the Storie Index is perhaps the most widely used. Any rating method is only as good as the person doing the classification; nevertheless, such methods help the observer focus on specific points and increase the consistency of evaluation.

The application of the Storie Index (Soil Conservation Service 1973:92) to the Son el Bricki profiles gives overall estimated values in the range of 58 to 72. This is in the lower range of Grade 2 and upper range of Grade 3 soils. Grade 2 soils are suitable for most crops capable of production within a specified climatic zone. This can be adjusted to include irrigation or nonirrigation as the case may warrant. Grade 2 soils have few management problems. Grade 3 soils are suited to fewer crops or to special crops and require special management. The Storie Rating is based on four factors: Factor A—profile characteristics in terms of relative suitability for plant root growth. Soils with deep permeable profiles would be rated at 100 percent. All profiles at Son el Bricki would rate 100 in this factor. Factor B—texture of surface soil as it pertains to ease of tillage and water-holding capacity. Moderate coarse- to medium-textured soils such as fine sandy loam rate 100. Profile 1, because of its shallow Ak horizon (0 to 5 cm) and immediate transition to sand that has low water-holding capacity, ranked the lowest among the four profiles. Factor C—slope, affects all profiles about the same. A value of 90 percent was given. Factor X—covers a range of conditions: poor drainage, high water table, erosion, acidity, salts, alkalinity, low fertility, and unfavorable micro-relief. Most of these problems are not serious at Son el Bricki, and fortunately, the micro-relief problems are easily alleviated by simple leveling.

The Entisols of Son el Bricki exist in a desert landform zone. The major physical factors which limit the resource use of these soils include low precipitation, low water-holding capacity, low fertility, and wind or water erosion hazards. Even so, with irrigation and good management practices these soils can be made productive in a sustainable agricultural ecosystem.

Aridisols

Aridisols are the soils that develop in stable topographic environments under arid climatic conditions. They are divided into two suborders—Argids and Orthids. Both are widespread in Sinai. Argids are desert soils which have a significant concentration of clay in a subsurface horizon. Within the Argid suborder there are five great groups. These groups are divided into subgroups based on moisture regimes, shrink-swell characteristics, rockiness, etc. Subgroups are divided into families depending upon particle size, mineralogy, and temperature. The occurrence of Argids is in such small parcels that they do not appear on small-scale maps such as Figure 5-1. Orthids, "true" Aridisols, are divided into six great groups based on profile characteristics. Like the great groups of Argids, these are subdivided into subgroups and families.

The five most common great groups of the Orthid suborder are:

- Typic Calciorthids—Dominate Tih Plateau of central Sinai. They have a calcic horizon (an accumulation of ≥ 15 percent

CaCO$_3$ in the B or C horizon) within 1 m of the surface. They are also calcareous (effervesces with cold dilute HCl) in all parts above the calcic horizon. These soils dominate about 32 percent of Sinai's soil surfaces.

- Lithic Camborthids—These desert soils are not saturated with water for ninety consecutive days or more within 1 m of the surface in most years. They have a lithic contact (bedrock) within 50 cm of the surface. They have a cambic horizon (subsurface diagnostic) characterized by somewhat altered materials (but not to the degree of an argillic or natric horizon). They have < 0.6 percent organic carbon, a texture finer than very fine sand, and less than 50 percent rock by volume. They have enough clay, usually 2:1 lattice types, to give an exchange capacity of 16 meq or more (fairly high for desert soils). Carbonate removal is evident (compare with Calciorthids above). This horizon shows stronger chroma and redder hue than underlying horizons. About 12 percent of Sinai's soil surface is dominated by this group.
- Salorthids—Salty desert soils with a salic (salty) horizon whose upper boundary is within 75 cm of the soil surface. These soils are saturated with water within 1 m of the surface for one month or more in most years. They have no duripan within 1 m of the surface. The soils cover about 4 percent of Sinai's surface and are limited to the northwest sector, where saline waters of the Mediterranean permeate the loose sediments of recent geological deposition.
- Typic Camborthids—Aridisols similar to Lithic Camborthids, discussed above, but having no lithic contact (bedrock) within 50 cm of the surface. These dominate about 1 percent of the soil area.
- Lithic Calciorthids—Similar to Typic Calciorthids (above), except they have a lithic contact (bedrock) within 50 cm of the surface. Found to be dominant on about 1 percent of the soil area.

Sinai has the geologic and geomorphic diversity to produce all the great soil groups found within the Argid and Orthid suborders. Due to the lack of detailed soil surveys, it is unknown if all great groups are represented and which great groups dominate. Reconnaissance surveys seem to indicate that Durorthids (those having hardpans in the profile) might be the most widespread group among the Aridisols.

As previously noted, the soil order of Entisols dominates northeast Sinai, but it was only after extensive field observation that the

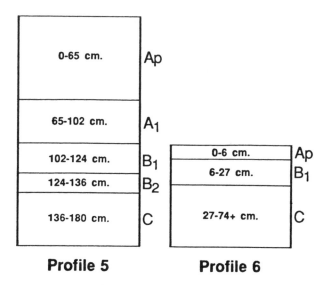

Profile 5　　　**Profile 6**

Figure 5-6. Aridisol profiles from northeast Sinai. Profile 5 is a Typic Durargid with a well-developed argillic (clay) horizon and an indurated (hard) layer throughout the ped. Profile 6 is a Typic Haplargid, similar to the Durargid but lacking the argillic horizon or an indurated layer within 1 m of the surface. Profile details are given in Table 5-2.

overwhelming dominance of Entisols was realized. The processes of geomorphology and pedogenesis are dominated by the forces of erosion and deposition. Much of the landscape is in slow but constant movement, so that it is hard to find sites where Aridisols have developed.

Profiles

Scores of sites were examined, of which only three gave the physical appearance of Aridisols, and were carefully sampled. Profiles 5 and 6, Figure 5-6, proved to be Aridisols. Table 5-2 details the results of laboratory analysis, color notations, and structural observations.

Profile 5 is the best example of an Aridisol found during the course of work in northeast Sinai. It has a well-developed argillic (clay) horizon above an indurated layer (hardpan) at the top of the C horizon. In most places the indurated layer is thick and massive. The profile has developed on coarse alluvium in an undulating

Table 5-2. Laboratory Analysis and Munsell Color Notations for Aridisol Profiles from Wadi el Arish Drainage

Profile Number	Horizon	Sample Depth (cm)	EC mmohs/cm	Total Soluble Salts (ppm)	Ph	Mechanical Analysis Clay %	Silt %	Sand %	Cation meq/1 Ca++	Mg++	Na+	K+	Munsell Color Dry	Moist
5	Ap	0–65	0.769	492.2	8.15	21.95	10.63	67.42	1.0	0.8	4.9	0.3	10 YR 8/4	7.5 YR 6/6
	A1	65–102	0.770	492.2	8.25	21.95	6.63	71.44	0.8	0.6	5.0	0.2	7.5 YR 7/4	7.5 YR 5/8
	B1	102–124	0.745	476.8	7.92	41.96	15.61	42.43	0.8	0.7	5.7	0.3	7.5 YR 6/6	7.5 YR 5/6
	B2	124–136	0.755	483.2	7.90	51.94	15.61	32.45	0.6	0.5	6.2	0.3	7.5 YR 7/4	7.5 YR 5/8
	B3	137–150	0.760	486.4	8.06	41.93	14.64	43.43	0.8	0.8	5.5	0.5	10 YR 7/4	7.5 YR 6/4
	C	150–180	0.772	494.1	8.17	41.50	25.68	32.42	0.6	0.4	6.4	0.3	10 YR 8/4	10 YR 7/6
6	Ap	0–6	0.716	458.2	7.40	26.95	15.61	57.44	0.6	0.5	5.7	0.4	10 YR 8/4	7.5 YR 5/5
	B1	0–27	0.693	443.5	7.35	12.92	32.60	54.48	0.6	0.6	5.4	0.3	7.5 YR 5/6	
	C	27–23+	0.699	447.4	7.55	11.91	23.63	64.46	0.6	0.6	5.5	0.3	10 YR 7/6	
7*	Cm	55–116+	0.213	501.1	7.68	6.90	25.65	67.45	0.4	0.3	6.9	0.2	10 YR 7/6	

*Note: in Profile 7 only the 55–116 cm layer was sampled. The top two layers were almost identical to those of Profile 6.

valley plain. It is well drained and appears to be moderately permeable.

The clay content for all horizons of profile 5 is considerably higher than for any horizon in the Entisols. Unfortunately total cation exchange capacities were not tested, but they are expected to be significantly higher than the 6–9 meq values found at Son el Bricki. The exchange capacities for these types of Aridisols are typically between 15 and 35 meq.

Profile 6 is not as definitive as profile 5. Horizon development is less differentiated. It is similar in color and structure to profile 5, but lacks an argillic horizon and an indurated layer within 1 m of the surface. The Ap (disturbed) horizon of this profile is an excellent example of an ochric epipedon (light-colored surface horizon with no organic carbon darkening). The processes of pedogenesis are not as fully developed here as in profile 5.

Work by J. Dan and associates on the soils of the Tih Plateau revealed profiles which in some cases were a little more complex than the Aridisols of the northeast. This is probably due to reg surfaces (gravels, frequently cemented by desert varnish), which provided a high degree of protection from both water and wind erosion. Dan classed these soils as "Regs," which correlates with specific Aridisols in *Soil Taxonomy*. Resource-wise, they are quite similar to the profiles discussed above, but have slightly shorter growing seasons.

Management

Management of the Aridisols of Sinai for agriculture follows the same general patterns as outlined above for Entisols. Slightly higher exchange capacities provide for better retention of nutrient elements. Hardpans may require ripping (deep plowing), and the application of irrigation will need careful planning for drainage. Lack of organic matter and relatively low cation exchange capacities will pose problems for fertility maintenance, as they do in all irrigated agriculture. Crop rotation and animal manures are the most important options. The long-range sustainability of desert agriculture is always problematic. Great care in planning and management is more crucial here than in most agriculture of the middle latitudes; nevertheless, it is possible to sustain a long-range human ecosystem in these soils.

The Need for a Comprehensive Survey

The movement of water from the Delta to the Tenna Plain (Salaam Project—see Chapter 7) is being proposed without adequate soil surveys to indicate the best areas to use this valuable resource. Governor Shaesh and his resource advisors like Hamdy Kotab of the Sinai Water Resource Agency are keenly aware of the pitfalls of proposed development without adequate background knowledge. The present proposal to dump a lot of water into the soils of north-west Sinai because they are closer to Nile water and lie below the 1 m contour, which will minimize the amount of pumping required, is indeed short-sighted. Unfortunately the lack of detailed soil information leaves opponents with inadequate response in terms of better alternatives. A detailed soil survey is sorely needed.

Based on the urgency of information need and general information gathered from scattered sampling and less-than-adequate reconnaissance mapping, it seems that the best alternative would be selective second-order mapping. The level of detail offered by a second-order survey would be adequate for areas dominated by Xerofluvents. A lesser detailed, more affordable third-order coverage could be used to cover most of the Aridisols and Torripsamments (Figure 5-1). Unfortunately surveys would probably cost over $2 per acre for second-order and $1 for third-order coverage. Depending upon acreage, such a survey could cost well over a million dollars. This sounds excessive, yet compared to the $48 million proposed by the World Bank for the Salaam Project and the potential waste that could result from putting that water in the wrong place, a soil survey could be a real bargain.

6. Biogeography of Sinai

In the spring of 1989 I was privileged to spend several days with Mohammed Kassas studying ecological patterns in North Sinai. Dr. Kassas, Professor Emeritus of Applied Botany at Cairo University, is a highly qualified scientist and Egypt's most widely recognized ecologist. He is an expert on the problems of desertification, an advisor to the United Nations Environment Program, past President of the International Union for the Conservation of Nature, and a consultant to Prince Philip of Great Britain on matters of international environmental deterioration. It was indeed a privilege to watch and work with this great scientist as he explained the intricacies of living communities and complex interactions of symbionts in seemingly sterile areas of Sinai's desert mountains and dune sheets. To Professor Kassas I owe a debt of gratitude for an appreciation of Sinai ecology and a greater understanding of its subtle workings and ability to persevere in the face of extreme environmental adversity.

In the harsh environments of Sinai both plants and animals must continually struggle just to exist. That struggle, however, has produced some very interesting associations and communities of life, plus a rich but scattered speciation in the plant sector and a relative paucity of animal species compared to most North American deserts.

Floristic Regions

The plant assemblages presently occupying the various regions of the earth are, for convenience, divided into floristic kingdoms, then subdivided into floristic regions. At present the most widely used system is based upon Engler (1924). In this system Sinai is split between the Boreal Kingdom (North Egypt and Syria Region) and the Palaeotropical Kingdom (North African–Indian Desert Region). This split between broad classification units hints at the complexities of flora that dominate the more specific levels of Sinai's plant geography.

Endemism and Phytogeography

While there are no precise definitions associated with the areal limits of plant endemism, it remains a highly useful concept in the study of phytogeography. An endemic organism may be visualized as one associated with a particular situation—a given region or part thereof. While endemism may apply to various taxonomic levels such as families, genera, or species, it is most often applied at the species level, which Good (1947:48) characterized as having an "abnormally restricted range." Endemism is especially useful in the differentiation of phytogeographic regions, discussed as "chorotypes." Within Sinai four such divisions exist:

1. Saharo-Arabian Species—most prominent in the sand deserts and hammada-covered plateaus of north and central Sinai, which receive less than 80 mm of rainfall,
2. Irano-Turanian Species—well developed in the higher plateaus and mountains, which receive 80–300 mm of precipitation,
3. Sudanese Species—tropical plants that have adapted to the dry conditions of the warm lowlands along the gulfs of Suez and Aqaba, which are on the warm side of the 23°C annual isohyte, and
4. Mediterranean Species—the least widespread and usually associated with a few small, favorable habitats within the Mediterranean Littoral or on the flanks of the Insular Massifs.

Environmental factors and floristic characteristics existing in Sinai were mostly developed during the Tertiary Period (Table 2-1), being strongly influenced by the existence of the Tethys Sea and the consequent separation from Eurasian floristic contact (Figure 2-1). The later influences of the Quaternary, in which ice greatly altered the floral characteristics of Europe almost to the Mediterranean Sea, did little to change the flora of Sinai.

Geological Factors

The closure of the Tethys Sea in late Miocene and the resultant land bridge between Asia and the Afro-Arabian area should have triggered a heavy floristic invasion of Sinai from the northeast. This failed to materialize due to increased aridity. Only in conjunction with certain upland and mountain genera (Irano-Turanian) is this potential noticeable. Otherwise, distributional patterns inherited from the Tertiary largely carried over into the present. Desert communities and genera dominate, while Mediterranean plants had limited impact on present distributions. Sinai is a crossroads in all ways. Furthermore, it is an area typical of convergent evolution, where lifeforms change and evolve to meet the stresses of a hostile desert environment. Undoubtedly many of the endemic species actually evolved there. Likewise, species which are presently nonendemic might well have evolved at this crossroad only to flourish elsewhere.

The Plants

The phytogeography of Sinai is rich in both variety and lore. Like many ancient desert formations, it is widely speciated. In evaluating the works of Hart, Post, Zohary, and Danin in terms of the *International Code of Botanical Nomenclature* (1988), the author estimated that Sinai has some 50 orders, 97 families, 472 genera, and perhaps as many as 1,224 species (Table 6-1).

Of the wide-ranging plants, the so-called "cosmopolitan" types, there are 130 genera worldwide. From this group, there are 31 really large genera (containing more than 50 species each). Of these, 16 are represented in Sinai. The largest is *Euphorbia*, with 20 species (worldwide it has some 1,000 species). Other important genera from this group include: *Plantago* with 17 species, *Salvia* with 8, and *Galium* with 7. Among the cosmopolitan genera having fewer than 50 species, Sinai is not well represented, having only 2 out of the 16 usually listed.

The next level of widely distributed genera, "subcosmopolitan," is generally represented by 80 genera worldwide, of which Sinai has 15, including 3 from the grass family. "Pantropical" genera are the next most widespread, including about 250 genera worldwide. Of this group 50 have more than 100 species each. Sinai is not well represented here since the vast desert barriers have kept out many plants originating in the moist tropics. Only 5 genera are significant: *Hibiscus, Ficus, Veronica, Indigofera,* and *Acacia* (Figure 6-1). The latter is by far the most important and best adapted to the intense aridity. It also includes the "shittim" wood, which the Bible says was used in construction of Noah's ark.

Most of the widespread temperate genera are missing. Only those with significant southern extensions such as *Artemisia* (sagebrush), *Rosa,* and *Prunus* are important. The Africa-wide groups, comprising 200–300 genera, have only a few examples in Sinai.

Figure 6-1. *Acacia raddiana* adjacent to Bedouin barley in the interior basin of Gebel Maghara.

Asia-wide groups, with 350–400 genera, have a similar paucity, an estimated 25 genera being represented.

More restricted in their worldwide distribution, but highly important to Sinai, are the regional groupings. Sinaitic regionals fall into two main categories: desert and upland-mountain types. Within these two are found most endemic and relict forms. Relict forms are persistent remnants of otherwise extinct genera such as the *Juniperus phoenicea*, which thrives in the uplands of the Insular Massifs. The 34 species now considered endemic are mostly found in the subdesert formations of the peripheral mountains or the higher mountain cores. Typical of these are ones derived from small flowering plants like pinks (*Caryophylleae*) and sunflowers (*Compositae*). Few endemics are found in the true desert plains and lower plateaus.

In addition to the endemic species, another 23 have only slightly greater distribution, limited to adjacent areas of the Negev or nearby parts of Jordan or Egypt. Regionally limited species, perhaps even more than widespread types, reflect Sinai's closer botanical ties with the East than with Egypt and the West. Regional desert species make up the largest single block of Sinai flora. At least 177 species in 131 genera have extensive distribution across the peninsula and into adjacent areas of climatic similarity in Egypt, Israel, Jordan, and Saudi Arabia. More than 60 species have significant

Table 6-1. *Higher Plants in Sinai*

TOTAL*

6 classes	50 orders	472 genera
10 subclasses	97 families	1,224 species (74 var.)

EMBRYOBIONTA (SPERMATOPHYTA) Seed Plants

4 classes	48 orders	467 genera
10 subclasses	94 families	1,220 species (66 var.)

MAGNOLIOPHYTA (ANGIOSPERMAE) Flowering Plants

2 classes	46 orders	466 genera
10 subclasses	92 families	1,216 species (63 var.)

(Worldwide 12,500 genera, 200,000 species)
MAGNOLIOPSIDA (DICOTYLEDONS) CLASS

6 subclasses	34 orders	381 genera
	73 families	999 species

(Worldwide 64 orders, 318 families, 170,000 species)

Magnoliidae Subclass—Woody Plants, Herbs, Forbs

2 orders	13 genera
7 families	30 species (3 var.)

(Worldwide 39 families, 12,000 species)

Hamaelidae Subclass—Nettles

1 order	3 genera
2 families	6 species

(Worldwide 11 orders, 24 families, 3,400 species)

Caryophyllidae Subclass—Pigweeds, Saltbush, Buckwheat

3 orders	59 genera
8 families	158 species (9 var.)

(Worldwide 3 orders, 14 families, 11,000 species)

Dilleniidae Subclass—Willows, Tamarisks, Violets

11 orders	79 genera
18 families	184 species (7 var.)

(Worldwide 13 orders, 78 families, 25,000 species)

Rosidae Subclass—Roses, Legumes

9 orders	82 genera
17 families	249 species (8 var.)

(Worldwide 14 orders, 114 families, 60,000 species)

Asteridae Subclass—Asters, Morning glory, Bindweed

8 orders	145 genera
21 families	372 species (22 var.)

(Worldwide 11 orders, 49 families, 60,000 species)
LILIOPSIDA (MONOCOTYLEDONS) CLASS

4 subclasses	12 orders	85 genera
	19 families	217 species (23 var.)

(Worldwide 19 orders, 65 families, 50,000 species)

(continued)

Table 6-1. (continued)

Alismatidae Subclass—Water Plantain, Frogbit
  3 orders   7 genera
  5 families   13 species (2 var.)
  (Worldwide 4 orders, 16 families, 500 species)
Arecidae Subclass—Palms, Cattails
  3 orders   5 genera
  3 families   6 species
  (Worldwide 4 orders, 5 families, 5,700 species)
Commelinidae Subclass—Rushes, Grasses
  3 orders   55 genera
  5 families   145 species (17 var.)
  (49 genera, 123 species, and 12 varieties of these are
    from the grass family [*Graminaea*])
  (Worldwide 7 orders, 16 families, 15,000 species)
Lilidae Subclass—Lilies, Irises
  3 orders   18 genera
  6 families   53 species (4 var.)
  (Worldwide 5 orders, 19–34 families, 25,000 species)
PINOPHYTA (GYMNOSPERMAE)—Naked Seed Plants
2 classes   2 orders   2 genera
             2 families   4 species
(Worldwide 3 classes, 400–450 species)
PINATA (CONIFEROPSIDA) CLASS—Pines, Junipers
  1 order   1 genus
  1 family   1 species
  (Worldwide 5 orders, 215–250 species)
GNETOPSIDA CLASS—Shrubs (Ephedras), Lianas
  1 order   1 genus
  1 family   3 species
  (Worldwide 3 orders, 66–71 species)
EQUISETOPHYTA (SPHENOPSIDA)—Horsetails
  1 order   1 genus
  1 family   1 species
  (Worldwide 1 genus, 15 species)
FILICOPHYST—Ferns
  1 order   3 genera
  2 families   3 species
  (Worldwide 28 families, 300 genera, 10,000–15,000 species)

*Estimates of Sinai plant speciation vary greatly. Most fall
between 900 and 1,200 species, but when a species by species
comparison is made between the extensive checklists of Post,
Zohary, and Danin over 1,200 species were found.

extensions both east and west, but most tend to be associated with one or the other directional grouping, i.e., found in Sinai and areas eastward but not westward and vice versa.

Upland and mountain plant varieties are perhaps the most interesting of the regional flora. A few show relationships to the poorly represented Mediterranean types, but most are related to species found in the elevated areas of Syria and Iran.

Mediterranean-wide species are limited because they are not well equipped to withstand the extreme aridity of most habitats, nor the winter cold of a few others. Great numbers of rodents also help check the spread of Mediterranean types, which tend to have few protective mechanisms. Nonviscous, nonnauseating, and sweet-tasting plants stand up poorly to rapid reproducers such as the common rat (*Rattus norvegicus*) and jerboa (*Jaculus*). In spite of this a few Mediterranean species like camel thorn (*Alhagi maurorum*) have thrived and become important economically as well as botanically.

Manna

Of the various plant foods of Sinai, none fired the imagination of European explorers and Bible scholars like manna, the "bread of heaven." Speculation is rampant on what it really was or is. The term itself has no exact meaning, and the biblical references are far from definitive. Theories fall along two main lines: "wind-blown" (lichens) and "honey-dew" (gum resins). Most theorists settle for the latter.

Many woody plants exude sap when the outer bark is broken. Frequently the exudates contain useful substances. Latex and maple syrup are examples of such resources. The manna theory has at various times included plants as diverse as legumes, willows, cedars, pears, eucalyptus, and tamarisk.

In Sinai, it is the honey-dew manna obtained from desert shrubs that receives the greatest attention. Most tamarisks, some acacias, and even camel thorn produce exudates. The focus, however, is on the *Tamarix mannifera* (tarfa), also rendered *T. gallica mannifera* (French tamarisk). The exudate production of all other shrub species combined is probably less than that of the tarfa. Both volume and quality set it apart.

Carl Ritter, the noted German geographer, in his exhaustive 1866 study of Sinai, devoted twenty-four pages plus numerous indirect references to touting tarfa as the biblical manna. While other scholars may not agree with Ritter, tarfa seems to be the logical candidate. Tarfa manna has for centuries been collected by Bedouin for their own diet and to sell to the pilgrims visiting the Judeo-Christian shrines of the peninsula. Distribution of *Tamarix mannifera* in Sinai is controlled by elevation and moisture; hence most occurrences are in wadi valleys below the 3,000-foot (914-m) contour. Three fairly localized occurrences dominate Sinai: Wadi Sheikh, Wadi Feiran, and Wadi Gharandal. Wadi Sheikh is a tributary of Wadi Feiran. For part of its course, just above the juncture with Feiran, it is known as Wadi Tarfa. This concentration around Wadi Sheikh is conveniently close to Gebel Musa (widely accepted as Mount Sinai) and Er Rahah (Plain of the Promulgation of the Law).

In addition to tarfa manna, other gummy exudates have over the years yielded economic value to Sinai. Gum Arabic, derived from acacias, is used in perfumes, medicines, candies, mucilage, and textiles. The best-quality gum is taken from *Acacia senegal*, not found in Sinai, but *Acacia arabica* yields a similar though somewhat inferior gum, which has periodically found a market when other supplies were poorly available.

Vegetative Communities

The evolution of vegetative communities is a complex process. At the broadest level it is controlled by macroclimatic factors—radiation, temperature, and moisture. Aridity by default becomes the prime factor influencing the formation of Sinai vegetation. But aridity is not solely a climatic factor.

Microclimatic factors, especially shade, water retention, and water movement, are relatively more important in dry climates. Likewise, deserts tend to retard the migration of plant species, thereby controlling community composition. The types of animals present, especially primary consumers, may further alter the nature of communities within similar climatic regimes. The influence of rodents, as with rats and jerboas, mentioned above, has contributed to the paucity of Mediterranean species in Sinai's communities.

The intricacies of Sinai vegetation are not widely appreciated, because of the usual sparsity of plant cover. The tendency is to perceive humid middle-latitude environments as having greater speciation because of their higher levels of ground cover. Danin (1983: 75) evaluated speciation for Sinai compared to other regions of the world based on a total of 812 species (an extremely conservative number considering the 1,224 species mentioned earlier). He calculated the concentration of species in Sinai at $21/km^2$, compared to: Sahara—20, Israeli deserts—50, California coasts—276, and the British Isles—125. From these numbers it is obvious that moisture is the primary limiting factor, but moisture availability is based upon a complex relationship among climate, topography, and soil. The more complex the geology and soil, the greater the diversity of water regimes and speciation within a given macroclimate. Thus an area containing many small geomorphological structures and varied soils will carry a more diverse vegetation than one dominated by large structures such as a single anticline and uniform soils. The total number of species in Sinai's deserts is further influenced by the steepness of climatic gradients (often a factor of abrupt elevational change) and the interaction of several phytogeographic regions (chorotypes).

Danin then evaluates speciation as a reflection of geomorphic and precipitation patterns. This he calls the "z" factor. A strong z value indicates a pronounced increase in the number of species with increasing area. His evaluation based on the z factor shows a far different pattern from that given above. Sinai now leads in diversity with a value of 0.307, followed by: Israeli deserts—0.299, Sahara—0.221, British Isles—0.209, and California coast—0.158.

Figure 6-2 shows Sinai's vegetative distribution organized in a way that reflects the interaction of climate, soils, topography, salinity, and isolation. The ten main vegetative communities (in bold) have been divided into four broad groups. Within the rich speciation of Sinai there are 33 species so well adapted or so unusual as to deserve detailed consideration. These are discussed in conjunction with the floristic group where they are most prominent.

Hot-Desert Communities of Plains and Low Plateaus (Accentuated, Hot Desert Climates, Köppen Class Mostly BWhl). **Desert scrub communities** of diffuse distribution, mostly reg (gravel) covers. These interfluve communities, which dominate the Tih Plateau and much of the Aqaba drainage, tend to have meager vegetative cover. The more important species include: *Anabasis articulata (Chenopodiaceae),* jointed anabasis, Bedouin "ajram," Saharo-Arabian chorotype, is one of those species which does well in several restricted environments. In the Dune Sheet and Dividing Valleys it has a diffuse distribution. South of the Dividing Valleys it is largely restricted to contracted distribution along the wadis. *Artemisia herba-alba (Compositae),* white wormwood, white sagebrush, Bedouin "shih," Irano-Turanian chorotype, has diffuse distribution in the Massif and Gebel Egma areas but limited to contracted patterns on the Tih Plateau. It is adapted to hard-rock areas as well as the reg surfaces found on the Tih. *Salsola tetrandra (Chenopodiaceae),* Bedouin "feers," Saharo-Arabian chorotype, is the most drought-resistant xerohalophyte, found in chalk and marl

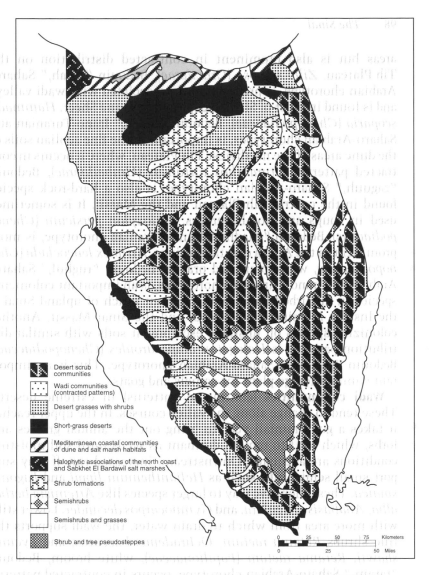

Figure 6-2. The vegetative groups of Sinai. Adapted from UNESCO 1970 and supplemented with data from Danin, Hart, and Zohary.
(1) *Hot Desert Communities of Plains and Low Plateaus* (Köppen Class Mostly BWh1): Desert scrub communities; Wadi communities; Desert grasses with shrubs; Short-grass deserts.
(2) *Subdeserts of the Mediterranean Coast* (Köppen Class Mostly BMh1): Mediterranean coastal communities of dune and salt marsh; Halophytic associations of the north coast and Sabkhet el Bardawil salt marshes.
(3) *Subdesert Uplands and Peripheral Mountains* (Köppen Class Mostly BWal and BWak): Shrub formations; Semishrubs; Semishrubs and grasses.
(4) *Subdesert Communities of the Mountain Cores of South Sinai* (Köppen Class Mostly GWB): Shrub and tree pseudosteppes.

Map legend:

Desert scrub communities
Wadi communities (contracted patterns)
Desert grasses with shrubs
Short-grass deserts
Mediterranean coastal communities of dune and salt marsh habitats
Halophytic associations of the north coast and Sabkhet El Bardawil salt marshes
Shrub formations
Semishrubs
Semishrubs and grasses
Shrub and tree pseudosteppes

0   25   50   75   Kilometers
0   25   50   Miles

areas but is also prominent in contracted distribution on the Tih Plateau. *Zilla spinosa* (*Cruciferae*), Bedouin "sillah," Saharo-Arabian chorotype, is predominantly associated with wadi valleys and is found in most of Sinai except for the Dune Sheet. *Hammada scoparia* (*Chenopodiaceae*), Bedouin "hidad," Irano-Turanian and Saharo-Arabian chorotypes, is important in the fine aeolian soils of the dune areas of the northwest Tih Plateau, where it occurs in contracted patterns. *Pituranthos tortuosus* (*Umbelliferae*), Bedouin "zaguuh," Saharo-Arabian chorotype, is also a hard-rock species found in the central Tih Plateau and the Massif. It is sometimes used in soups and hot beverages. *Haloxylon persicum* (*Chenopodiaceae*), Bedouin "ghadha," Irano-Turanian chorotype, is most prominent in the Aqaba coastal sections. *Atriplex leucoclada* (*Chenopodiaceae*), white-branched orache, Bedouin "rughul," Saharo-Arabian and Irano-Turanian chorotypes, is an important colonizing species in disturbed habitats throughout much of upland Sinai— the Insular Massifs, Tih Plateau, and the Sinai Massif. Another colonizing species on unstable escarpment soils with similar distribution patterns is *Halogeton alopecuroides* (*Chenopodiaceae*), Bedouin "sh'aran," Saharo-Arabian chorotype. This is an important summer-forage plant for camels and goats.

Wadi communities (contracted patterns) in extreme deserts. These tend to change along the valley courses. In the upper reaches it takes a good rainfall year to bring out the annual grasses and forbs, which tend to remain dormant as seeds until soil moisture conditions are adequate. Downstream, increased runoff may support small semishrubs such as *Helianthemum lippii* and *Fagonia sinaica*. These then give way to larger species like *Artemisia herba-alba*, *Anabasis articulata*, and *Gymnocarpos decander*. Lower still, with more area from which to drain water, the wadi supports the true shrubs *Retama raetam*, *Ochradenus baccatus*, and *Lycium shawii*. *Retama raetam* (*Papilionaceae*), white broom, Bedouin "ratam," Saharo-Arabian chorotype, occurs in contracted patterns in the sandy parts of the vast basin of the Wadi el Arish, which drains the Tih Plateau. *Retama* is a "stem-assimilant"—the green stem does much of the photosynthesizing, especially during the dry season. Fruits and flowers provide forage for goats. The stems provide fuel directly and are also converted to charcoal. Finally, near the wadis' base level, with optimal water catchment, trees such as *Acacia raddiana*, *A. tortilis*, and *Tamarix nilotica* become dominant. *Stipa capensis* (*Gramineae*), twisted-awn feather grass, Bedouin "safsuf," Irano-Turanian and Saharo-Arabian chorotypes extending into Mediterranean, is the most important annual grass in Sinai. In rocky habitats of the Tih Plateau, the Insular Massifs,

Gebel Egma, and the higher parts of the Suez Foreshore, it is frequently associated with *Zygophyllum dumosum* (*Zygophyllaceae*), bushy bean caper, Bedouin "'adhbee," Saharo-Arabian chorotype (relative of the North American creosote bush). *Anastatica hierochuntica* (*Cruciferae*), rose of Jericho, Bedouin "kaff-e-rahman," Saharo-Arabian chorotype, is perhaps the most unusual annual in Sinai. The curved, compact stems of the dead plant, when immersed in water, will straighten, opening up like a living organism. Christian pilgrims considered this symbolic of the Resurrection—hence "anastasis," the Greek word for resurrection. To the followers of Islam, the opening of the fist-like dry plant into the open "hand of the merciful" (kaff-e-rahman) signifies the beneficial changes that follow substantial rainfall. In a similar vein the Muslims also call it "kaff-e-nabbi" (hand of the prophet).

Desert grasses with shrubs. These are important communities in Sinai, but great variety and areal distribution exist in mostly diffused patterns. The most important grass through this entire region is the perennial grass *Stipagrostis scoparia* (*Gramineae*), triple-awned grass, Bedouin "sabat," Saharo-Arabian chorotype. It is a dune dominant and often the exclusive species in the fine-grained sand of mobile dunes east of the Suez Canal and extending through the Insular Massifs to the area between El Arish and Gebel Halal. It can also exist on saline soils to the very edge of salt marshes. Dominance also extends southward in the Suez Foreshore and Plain of Qa. It is a true survivor in a variety of Sinai's toughest habitats. Its best survival technique is to grow near shifting dunes, where competition is slight and water availability is heightened by the insulative properties of the sand. Another perennial grass, *Panicum turgidum*, traps blowing sand in wadi bottoms, where it increases water infiltration and improves the edaphic environment and moisture retention. Within the same general climate zone, but on more stable surfaces, species dominance shifts to annual grasses mixed with *Anabasis articulata*, *Artemisia herba-alba*, *Zygophyllum album*, or *Thymelaea hirsuta* (*Thymelaeaceae*), shaggy sparrow-wort, Bedouin "mitnan," Mediterranean and Saharo-Arabian chorotypes. The latter is usually found in contracted distribution on the hard-rock habitats that sometimes outcrop in the dune areas. Avoided as a fodder plant, even by hardy black goats, its value to the Bedouin comes as a fiber for ropes and mesh for carrying water jugs, etc. Farther south along the Suez Foreshore and Plain of Qa the dominant semishrub is *Hammada salicornica* (*Chenopodiaceae*), Bedouin "rimth," Sudanian chorotype. Its chorotype is reflected in the fact that it does not occur at latitudes north of the Little Bitter Lake. *Astragalus camelorum* is a shrub endemic to this southern sector. In this same area *Retama raetam* exhibits great adaptability, being one of the few shrub forms that can survive dune depth greater than 1 m.

Of the tree forms found in Sinai, three species, the so-called "Sudanian trees," are usually considered among the most striking. Although they are not limited to the peninsular part, they are definitely more common from Suez and Aqaba southward along the coastal strips and extending into the Massif along the major valleys. Their tolerance for heat is manifest in their distribution. *Acacia raddiana* (*Mimosaceae*), Bedouin "sayal," Sudanian chorotype, will grow in elevations up to 1,400 m, but their predominance with increasing wadi size indicates a preference for more stable water regimes. *A. tortilis* is more often associated with smaller wadis. Both species have the ability to tap underground sources as deep as 12 m. The buds and fruits are important forage for sheep and goats. These acacias are not particularly dependent upon annual moisture conditions, making the species a highly reliable resource. The Bedouin, therefore, do not cut these trees for fuel. Gum secretions like those of *Tamarix mannifera* are sweet and used as a basis for candy. *Phoenix dactylifera* (*Palmaceae*), date palm, Bedouin "nakhlih," Sudanian chorotype extending into Saharo-Arabian, is one of the earliest domesticates, going back some 6,000 years. Wild date palms have many stems and small but tasty fruits. Wild trees are widely distributed in Sinai. Domestic varieties are numerous and widely cared for, including hand pollination. The date palm is dioecious, male and female flowers develop on separate trees. Wind is the usual mechanism of pollination. This is the most useful tree of the desert. Its fruit is an important food source. Palm leaves are used to build shelters and fences. Narrow strips are cut to make baskets. Hollowed trunks are used as water conduits for irrigation. As a fuel in the oasis areas it has no equal. *Ziziphus spina-christi* (*Rhamnaceae*), Syrian Christthorn, Bedouin "sidir," Sudanian chorotype extending into Saharo-Arabian, Mediterranean, and tropical African, is highly scattered in Sinai.

Short-grass deserts. An estimated 48 genera with over 134 species of *Poaceae* (grass family) are found in Sinai. Most of these are annuals. In spite of this wide speciation, it is doubtful if annual grasses dominate any community. It is again perennial grasses such as *Stipagrostis scoparia* and *Panicum turgidum*, mentioned in the previous section, that respond well to habitat pressures. Annuals like *Stipa tortilis* form vegetative communities with *Artemisia herba-alba* and *Crotalaria aegyptiaca* (*Leguminosae*) on the sand plains north of Gebel Maghara and southeast of El Arish. *Anabasis* species are also important to soil stabilization in this area. *Orobanche* (dodder or cancer root) exists in a parasitic symbiosis with the *Anabasis*. *Mesembryanthemum forsskalii* (*Aizoaceae*), similar to fig marigold, Bedouin "samkh," Sudanian chorotype extending into Saharo-Arabian, is an unusual succulent annual found mostly in the Aqaba Foreshore. The seeds are ground to make a not-so-tasty bread. Along the narrow plain and embayments of the Aqaba coast, grasses give way to salt marshes dominated by halophytes such as *Zygophyllum album, Nitraria retusa, Suaeda vermiculata,* and *Limonium axillare.* But most unusual, in muddy soils of dead coral reefs and tropical tidewater, from Ras Mohammed to Nuweiba, are the mangrove swamps formed by *Avicennia marina.* Stands of *Salvadora persica* (toothbrush tree) lend another dimension of variety to this area. Short-grass associations probably show the effects of millennia of sustained overgrazing more than any other habitats. In the oases of the short-grass deserts of the Aqaba Foreshore is found the *Hyphaene thebaica* (*Palmaceae*), doum palm, Bedouin "dom," Sudanian chorotype. It is cultivated by the Bedouin for its fruit, which is eaten fresh or dried and ground to make cakes.

Subdeserts of the Mediterranean Coast (Long Dry Season, Tendency for High Humidity, Köppen Class Mostly BMhl). **Mediterranean coastal communities of dune and salt marsh habitats.** Mostly diffused patterns. Desert grasses with shrubs *Stipagrostis scoparia* and *Anabasis articulata,* usually in exclusive stands, dominate the mobile dunes of fine-grained sand. In coarse gravel habitats, the *Artemisia monosperma–Thymelaea hirsuta* association is dominant. *Artemisia monosperma* (*Compositae*), single-seed wormwood or sagebrush, Bedouin "ʾadhir," Saharo-Arabian chorotype, is one of the species adaptable to sandy or gravelly habitats. In the eastern part of the Dune Sheet it is found in diffused distribution. In the western Dune Sheet and the Insular Massifs it tends to occur in contracted distribution in the wadis. *Cornulaca monacantha* (*Chenopodiaceae*) and *Convolvulus lanatus* (*Convolvulaceae*) tend to be found as exclusive stands in areas of wind-blown sand, their taproots being protected by thick bark. *Erodium hirtum* (*Geraniaceae*), hairy storks-bill, Bedouin "bilbis," Saharo-Arabian chorotype, is one of the few plants, along with *Stipagrostis scoparia,* able to survive in fine-grained, mobile dune areas, especially along the Mediterranean side of the Dune Sheet and the coastal plains of the Gulf of Suez.

Halophytic associations of the north coast and Sabkhet el Bardawil salt marshes. Here a variety of halophytic associations are dominated by *Halocnemum strobilaceum, Arthrocnemum macro-*

stachym, *Suaeda aegyptiaca*, and *Limoniastrum monopetalum*. In less saline marshes *Juncus arabicus* (rush) and *Phragmites australis* (common reed) become dominant. *Carex divisa* (sedge), *Scirpus maritimus* (tule), and *Typha angustata* (cattail) round out this uncharacteristic Sinai vegetative sequence. In the adjacent salty soils *Zygophyllum album* and *Meteria etosia* are important. The latter species was identified by Dr. Mohammed Kassas in the field but not verified in formal literature.

Subdesert Uplands and Peripheral Mountains (Long Dry Seasons, Cool Winters, Köppen Class Mostly BWal and BWak). **Shrub formations,** mostly diffused distribution dominated by *Artemisia herba-alba, Anabasis articulata, Zygophyllum dumosum,* and refugial communities of *Juniperus phoenicea* and infrequently *Pistacia khinjuk (Anacardiaceae)*, pistachio, Irano-Turanian chorotype. The latter grows in contracted patterns. *Zygophyllum dumosum* is well adapted to hard-rock surfaces, showing diffused distribution throughout the Insular Massifs and along the western edge of the Tih Plateau. *Juniperus phoenicea (Cupressaceae)*, Phoenician juniper, Bedouin "'ar'ar," Mediterranean chorotype, forms rare refugial stands in the Insular Massifs, where the paleobotanic record goes back 34,000 years in Gebel Maghara. Associated with the juniper are a number of diverse relict forms of the Mediterranean chorotype, including: *Astoma seselifolium (Umbelliferae), Pancratium parviflorum (Amaryllidaceae)* and *Rubia tenuifolia (Rubiaceae)*.

This unit includes areas where bedded limestones support six important vegetative associations such as *Zygophyllum dumosum–Reaumuria hirtella (Tamaricaceae)* and *Artemisia herba-alba–Noaea mucronata. Noaea mucronata (Chenopodiaceae),* thorny saltwort, Bedouin "sirr," Irano-Turanian chorotype, is a typical desert semishrub frequently associated with hard-rock areas. Smooth-faced limestone tends to support single dominants such as *Stachys aegyptiaca (Labiatae), Iphiona mucronata (Compositae),* and the refugial *Juniperus phoenicea.* Chalk and marl outcrops are often associated with exclusive species including *Suaeda vera, Reaumuria hirtella,* and *Salsola schweinfurthii.* Sandy habitats support exclusive species like *Hammada salicornica* and associations that tend to include *Artemisia monosperma. Salsola inermis (Chenopodiaceae),* Bedouin "khadhraf," Saharo-Arabian chorotype, is a halophytic summer-annual forb and sometime colonizer in disturbed soils that have not been depleted of winter moisture. It also occurs with *Zygophyllum dumosum* or *Artemisia herba-alba* on stony soils in the Insular Massifs, along the western side of the Tih Plateau and Gebel Egma.

Semishrubs dominated by *Salsola tetrandra, Halogeton alopecuroides, Artemisia herba-alba,* and *Hammada salicornica.* The foregoing are found in both diffused and contracted distribution along the wadi channels. This vegetative sequence is associated with upland chalk areas along the Gebel Egma and western edge of the Tih Plateau. In places the wadis lead down into the sandstones of the Dividing Valleys, which intensifies the contracted distribution of vegetation along the descending valleys. In sandy wadis the contracted sequences include: *Hammada salicornica, Retama raetam, Ephedra alata (Gnetaceae), Iphiona scabra, Anabasis articulata,* and *Zilla spinosa.* In silty areas the sequences include: *Salsola cyclophylla, Reaumuria hirtella, Artemisia herba-alba, Gymnocarpos decander,* and *Anabasis setifera. A. setifera (Chenopodiaceae),* Bedouin "gilu," Saharo-Arabian chorotype, is an important colonizing species in disturbed areas. It grows in contracted patterns in the southern two-thirds of Sinai.

Diffuse vegetation appears in a variety of formations frequently controlled by the nature of the substrate. In the areas of massive sandstone, persistent communities of *Launaea spinosa (Compositae), Pituranthos triradiatus (Umbelliferae),* and *Salsola cyclophylla* are common. The loose sand coverings mixed with other strata support a wide variety of communities, which include *Retama raetam* and *Heliotropium digynum (Boraginaceae), Ephedra alata* and *Convolvulus lanatus,* and areas of the single dominant *Haloxyletum persici. Noaea mucronata* is important in hard-rock areas along the Gebel Egma, where it shows a diffuse distribution. In the same hard-rock areas, but associated with a contracted pattern along the wadis, is *Varthemia iphionoides (Compositae),* Bedouin "sleemanee," an unusual Mediterranean chorotype extending into Irano-Turanian and Saharo-Arabian. *Centaurea eryngioides (Compositae),* Bedouin "digin-el-badan," Irano-Turanian chorotype like *V. iphionoides,* inhabits rock outcrops and cliff crevices. Its flower head resembles the *Cirsium ochrocentrum* (yellow-center thistle) of the American Colorado Plateau.

Semishrubs and grasses. The semishrubs are similar to the foregoing group but with less diversity and increasing grasses such as the perennial *Panicum turgidum.* Avinoam Danin includes much of this area with the Tih Plateau vegetation; however, the higher interfluves extending between the El Arish and Dead Sea drainages provide for greater coverage than either adjacent area. *Anabasis articulata* and *Anabasis setifera* are in places found as exclusive species. Contracted distribution in the smaller first- and second-order wadis tends toward *Artemisia herba-alba* and *Hammada scoparia.*

Subdesert Communities of the Mountain Cores of South Sinai (Shorter Dry Seasons, Cold Winters, Köppen Class Mostly GWB). **Shrub and tree pseudosteppes.** Danin divided the habitats into fissured rocks with stony soil, smooth-faced granite, and wide valleys. He lists eight associations in the first group, each including *Artemisia herba-alba* in association with other shrubs such as *Gymnocarpos decander* or *Anabasis setifera. Rhus tripartita (Anacardiaceae)*, Syrian sumac, Bedouin *"'irn,"* Irano-Turanian chorotype grading into Mediterranean and Saharo-Arabian, is most widely distributed in the Massif, but found in elevated portions of Gebel Egma and Gebel Maghara. It is similar to several shrubs in southern California brushlands.

Six associations are found in the smooth-faced granite, including *Globularia arabica–Verbascum decaisneum* and *Varthemia montana–Pistacia khinjuk.* In the wide valleys a single association, *Achillea fragrantissima (Compositae)–Artemisia judaica,* dominates. Near larger springs a variety of oasis species thrive— *Phoenix dactylifera, Tamarix nilotica, Mentha longifolia (Labiatae)*, and *Holoschoenus vulgaris (Cyperaceae)*, club rush. *Crataegus sinaica (Rosaceae)*, Sinai hawthorn, Irano-Turanian chorotype, is typical of the isolated species that occur in the coolest parts of the Massif. The nearest areas of wide distribution are in Iran and Lebanon. This tree produces an edible fruit, but the innovative Gebelia Bedouins near Gebel Musa graft pears onto its root stocks. Throughout the Massif and extending to the Suez and Aqaba coasts is found *Blepharis ciliaris (Acanthaceae)*, Bedouin "shok-e-dhab," Sudanian chorotype extending into Saharo-Arabian. This wadi species is one of the most widely illustrated Sinai plants. The Latin name means "eyelash," while the Arabic refers to "lizard tail." The plant is quite adaptable, being either an annual or perennial depending upon water availability.

The Animals

Limited and scattered biomass greatly restricts the animal populations of Sinai. Fewer species, smaller numbers, and a greater proportion of small-bodied creatures are the rule. As in most ecosystems, small creatures such as lizards, rodents, and birds are far more common than gazelles or leopards.

The precarious existence of desert fauna emphasizes certain

Figure 6-3. Common or three-toed jerboa, *Jaculus jaculus*, after Tristram.

characteristics—the running, jumping, and burrowing capabilities of many species seem to increase with a decrease in protective cover. Lizards and jerboas depend upon their speed and jumping ability for defense (Figure 6-3). The common jerboa (*Jaculus jaculus*), literally "jumping mouse," is a difficult target for falcons and jackals. Burrowing lizards can wiggle into the sand so quickly that most predators are unable to capture them. Burrowing reptiles also include the *Atractaspis engaddensis* (black mole viper), whose color-blending adaptation adds to the effectiveness of the burrowing.

Faunal Regions

The zoogeography of Sinai, like its phytogeography, has been extensively shaped by crustal tectonics. The closing of the Tethys Sea, accompanied by a process of global desiccation and the subsequent role of Sinai as a land bridge during all of the Quaternary Period, has shaped the recent patterns of animal distribution. According to the great zoogeographer Alfred Wallace (1876:180), Sinai falls completely within the Desert Belt of the Mediterranean Subregion of the Palearctic Region of Zoological Geography. Recent biogeographers, such as Peter Furley (1983), confirm the validity of this division, yet there remains a sizeable component of tropical species. Tropical species originating in the Ethiopian Region are found in a number of biocommunities, especially in the southern part of the peninsula.

The *Procavia capensis* (rock hyrax), of tropical derivation, together with the ibex, of the Palearctic, frequently dominates the rocky slope communities of the nearly inaccessible mountains of the Sinai Massif. Ironically the hyrax, sometimes misnamed "rock rabbit," is a hoofed ungulate, more closely related to ibex and gazelle than to its look-alikes among the rodents. Likewise, sand partridges and pigeons of the north are closely associated with the tropical *Cinnyris osaea* (sunbirds) and *Onychognathus tristrami* (Tristam's grackles) that inhabit the gorges and deep wadi channels around Zebir, Musa, and Serbal.

In recent geologic times, broad deserts closed off the exchange of terrestrial and freshwater species that had occurred extensively in Sinai prior to the Miocene. Increased biological isolation led to endemism and refugial communities. The previously mentioned sunbirds, grackles, and hyraxes were cut off from larger breeding groups and have developed various degrees of endemism, and all exist in essentially refugial communities. These changes are sometimes adaptive. In Sinai they commonly produced "xerotropic forms" that are better able to withstand the aridity and wide swings of temperature than their tropical cousins.

Mammals

Only two large-bodied, wild herbivores are found in Sinai at present. These are the Nubian ibex (*Capra ibex nubiana*) and Dorcas gazelle (*Gazella dorcas*). Gazelles, like most antelopes, are denizens of the open plain, relying upon speed and keen senses for their safety. They are able to survive on extremely poor fare and without regular intake of water. Even in the stress of summer heat an intake of nothing but acacia leaves often provides both food and water requirements for several days. They really thrive on the spring shoots of *Calligonum comosum* and *Rumex vesicarius* (both in the buckwheat family), which cover the sandy plains and low plateaus for a few weeks each year. This graceful but small member of the deer family has rather large lyre-shaped horns that often appear as a single horn when viewed in profile. It is thought, by some, to be the source of the unicorn legend. Livestock diseases and habitat competition from domestic herds add to direct predation in reducing their numbers. Distribution has become highly sporadic, and the gazelle has been extirpated from most of its former range. The small herds of the present cannot survive even a few predators, so continued survival will depend upon direct intervention on their behalf.

The ibex differs significantly from the gazelle. It must drink daily in hot weather, so its range is restricted by available water. In summer this means access to permanent springs, which drastically curtails its range. The seeming inaccessibility of its rugged mountain habitat is increasingly constricted by human encroachment. Like the gazelle, its survival depends upon human policy.

Birds

Few bird species reside in Sinai permanently. The real residents are predominantly ground birds with limited distance-flying capabilities and shore birds that exploit the rich sea margins. The *Struthio camelus* (ostrich), probably the only flightless bird of recent times in the peninsula, once roamed widely over the northern plains. Unfortunately it has been extinct in Sinai since the late 1800's. It was no match for the barren environment and a chief predator with firearms. Arms allowed the Bedouin to supply feathers and skins to markets in Beersheba and elsewhere. Sadly, more than a hundred years following their extinction, ostrich egg shells are still occasionally found in the dunes.

Other ground birds have long been important resources for Sinaitic peoples. The miracle food provided by quail for the Children of Israel is widely recounted. Only slightly less important to local tribesmen and early European explorers was the little bustard (*Tetrax tetrax*), which provided many a meal for hungry travelers. For modern Bedouin the sand partridge (*Ammoperdix heyi*), spotted sand grouse (*Pterocles senegallus*), snipe (*Gallinago gallinago*), and quail (*Coturnix coturnix*) still yield limited food resources. In the case of quail, the vast market in Europe, which provided an important currency source for Sinai prior to World War II, can no longer be supplied because of dwindling numbers. The critical zone of the breeding stock has been broached and recovery seems unlikely.

The real majesty of birds in Sinai, however, occurs during a few weeks each spring and fall. Sinai lies under the flyway for numerous Palearctic species that nest at high latitudes in summer but return to the tropics or subtropics for winter. Stragglers of some species like the white stork (*Ciconia ciconia*) may winter in Sinai, although most proceed south to the coast of east Africa.

Waterfowl of all sizes, small songsters and numerous birds of prey, which follow the moving feast, all contribute to the host. So varied

Figure 6-4. Nubian ibex, *Capra ibex nubiana*, after Tristram.

is the migration that Sinai entrepreneurs might gear some of their promotion of tourism to the times of annual movements.

Of the songsters and other small species the majority are migratory. Those species, found widely in Europe and western Asia in summer, are often found in Sinai in fall and spring. The actual movement of the small insectivores is usually nocturnal, whereas fast fliers like pigeons, crows, jays, and hawks are more apt to migrate during the day.

Among the less migratory species is the cream-colored courser (*Cursorius cursor*), which is well adapted to live in sandy deserts. Its long, curved bill is used to extract beetles from the sand. The brightly colored grackles and the sunbirds, both with strong blood ties to tropical Africa, also have restricted migratory habits. Sometimes the range is from mountains in summer to lowlands in winter.

The majority of water bird species are migratory, moving from their African wintering grounds to northern Europe and Asia for summer nesting. Ducks, grebes, and geese have rather short residencies. Their powerful flight is in either day or night. Cranes and egrets are mostly passage migrants that take advantage of good food or water for a few days, then move on. The western reef heron (*Egretta gularis*), however, has a more limited migratory range, nesting in the mangroves around Ras Mohammed and up the Aqaba shoreline (Figure 6-5).

Raptors and other birds of prey, especially accipiters, fast low-flying, short-winged hawks, prey on migratory species and tend to move along with them. Osprey and other fishing hawks are more apt to be permanent residents, depending upon marine resources. They nest on Tiran Island and adjacent areas of the Aqaba coast.

Birds here have adapted to the extremes of climate in various ways. The little owl (*Athene noctua*) is nocturnal, therefore somewhat insulated from the extremes of temperature; it can hunt its rodent and reptile prey year-round. The raven (*Corvus corax*) utilizes an omnivorous diet, but one food source is in rather constant supply, blood-bloated camel ticks, so ravens often form symbiotic relationships with the camel herds.

On the whole, Sinai's birds are a dynamic lot. The number of species constantly changes through the year. The greatest variety, however, is found in the spring and fall when migrations occur and weather conditions are best suited for birds and observers.

Reptiles and Amphibians

As might be expected, reptiles are far more common than amphibians. Six species of poisonous snakes are found here, plus several harmless varieties. The largest snake, the desert cobra (*Waltherinnesia aegyptia*), is also the most dangerous. It can grow to more than 5 feet and strike for a goodly portion of that length. Fortunately it is rare and does not spit its venom as do several of the Indian and African varieties. Most snakes are harmless to people and largely prey on small rodents and lizards.

PLOTUS LEVAILLANTII

Figure 6-5. Western reef heron, *Egretta gularis*, after Tristram.

Variety is the rule among Sinai lizards. The small whiptail lizards (*Lacertidae*) are the most common and widely speciated. The largest lizards in the Middle East, the desert monitor (*Varanus griseus*) and Egyptian spiny-tail agama (*Uromastix aegyptius*), both live in Sinai. Sand lizards (*Acanthodactylus boskianus*) show the greatest tolerance for arid climates and have special adaptations for running on loose soils—their toes are fitted with soft scale extensions that provide better footing. *Agama* species are noted for their preference for extremely high temperatures. As a group, lizards are probably Sinai's most highly adapted creatures.

Frogs and toads are significant only in favored spots where moisture softens the soil and stimulates a luxuriant plant cover, which usually includes some frogbit species (*Hydrocharis*). In spite of their rarity, the noise of courting frogs can rouse the ire of the Bedouin.

Insects and Arachnida

From the great diversity of the classes *Insecta* and *Arachnida* only the more prominent will be discussed. One of the most widely speciated orders found here is the beetles (*Coleoptera*). More than 143 species are found, of which an estimated 26 are endemic. Frequently, such species as the small beetle (*Adesmia acis*) are important to several food chains that support mammals, birds, and reptiles. For example, the previously mentioned courser thrives on beetles extracted from loose sands by its specially adapted beak.

Diptera, the true or two-winged fly, is another order of wide speciation. It is represented by many species of stumpy flies with maggot larvae (*Cyclorrhaph*), which include houseflies and a wide range of parasitic biters similar to deerflies. Mosquitoes, especially *Aedes aegypti* (yellow fever vector) and *Anopheles* species (malaria vector), continue to be health factors. In general flies and mosquitoes are a greater problem on the plains than in the high mountains, where winter cold reduces the breeding season.

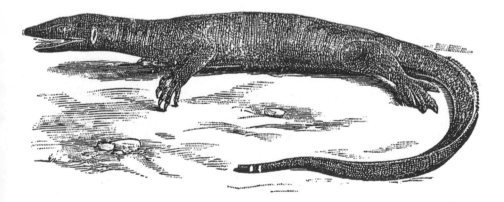

Figure 6-6. Desert monitor, *Varanus griseus*, after Rawlinson.

The order of ant, bee, wasp, and hornet (*Hymenopter*) is amply represented, but with a reversal from the typical dominance in that hornets are relatively more abundant than wasps. As might be expected, ants have close relationships to a variety of plants, the most interesting being the *Tamarix mannifera*, which supplies them with the same nectar supposedly received by the Host of Israel.

Sinai is not the butterfly collector's paradise. The species most frequently seen tend to be the smaller and less gaudy varieties such as the brimstone (*Gonepteryx rhamni*) and the small heath (*Coenonympha pamphilus*), which are migratory from Europe. The heat of the "terrible summer" is too much for them.

Cicadas, like butterflies, are mentioned here more for their appeal than any large contributions they make to the ecology. The song of the adult male is probably the loudest sound made by any insect. The *Tibicen* species, with a yearly reproductive cycle, is most common. They are sometimes called "dog days cicadas" because of their appearance at midsummer. *Magicicada* species have a long underground nymph state, sometimes lasting several years. This appears to have adaptive survival value in the aridity of Sinai. Cicadas have been a source of human food in different cultures, including some in Sinai, where Greek influences may have been the root. They are also kept as pets by some young Bedouin.

The order of grasshopper, locust, cricket, katydid, and mantid (*Orthoptera*) is certainly the most significant insect group in Sinai ecology. Except for the manitids and cockroaches, this order is largely herbivorous. The large, plant-eating grasshoppers, noted for their oversized heads, strong jaws, and sharp teeth, are probably the peninsula's greatest bane. About a dozen of the grasshopper species found in Africa and the Middle East are called "locusts" because of their strong migratory habits. Desert locusts (*Schistocerca gregaria*) and migratory locusts (*Locusta migratoria*), with various subspecies, are the most common insects found in Sinai. Outbreaks of these destructive creatures usually accompany irregularities in precipitation and soil moisture or periodic grass fires. For Sinai, it is external events which trigger most outbreaks. Suitable soil moisture and warmth in parts of Africa stimulate rounds of breeding and hatching that force an outward migration. Distances traveled can be great. Locusts from eastern Africa have no trouble reaching Sinai and devastating its vegetative cover, poor though it may be. Perhaps the greatest locust swarm ever noted by man was one of desert locusts that crossed the Red Sea in 1889. It was estimated to cover an extent of 2,000 square miles (5,180 km^2) as it moved across the Red Sea to overwhelm Sinai in the process. Mortality rates in such a mass are high, but not rapid enough to stave off equally rapid reproduction and the wide destruction of almost every living plant and many other objects in its path.

So it goes, with wide swings of biological activity corresponding to the changing climatic phenomena. High aridity and insolation in particular influence the nature and the fluctuations of Sinai's ecosystems and the human use thereof. The sparsity of biomass in most communities has a direct impact on the types and number of animals in Sinai's ecosystems. Resident populations of mammals are small in both numbers and species. Most of the predators like wolves and hyenas are all but gone. Smaller predators like the red fox (*Vulpes vulpes*) and Ruppels fox (*Vulpes ruppelli*) are a bit better suited to withstand the encroachments of mankind. Small populations of ibex and gazelle have survived. Small rodents and burrowing animals like the hyrax, and jumping rodents like the jerboa, have fared better. Birds continue to be important to Sinai's ecology. They reflect a greater ability to withstand the high aridity and intense insolation of this harsh environment.

The Human Ecology

The swings of biological activity and the slow but constant decline of biological diversity in Sinai parallel the changes in the fragile human ecosystems existing there. The successes and failures of individual species are but microcosms of Sinai's greater ecosystem. The obvious impact of mankind on physical habitats, lifeforms, and biological communities is at an all-time high. The scope of human migration from Egypt's crowded valleys and Delta cannot be ignored. The ultimate impact of human society with its population growth and increased mechanized mobility is bringing this fragile land to a more crucial turning point than it has ever known in the long millennia of cultural change. Some in Sinai's government are aware of these problems, and we can only hope they are strong enough and wise enough to mitigate the course of human dominance that continues to sweep away the great biological heritage of this unique land. In the next and final chapter we will examine the problems of sustainability and will suggest a few alternatives for the inevitable human intervention.

7. May They Eat Lamb in Paradise

For Sinai in general and among the Bedouin in particular there is much uneasiness about the future. Consciously or subconsciously there is a feeling that they must move toward a sustainable future. National policies for the resettlement of peasants from the heavily populated lands of the Delta and Nile Valley, combined with high levels of natural population increase, are viewed with understandable misgiving. Unfortunately the cushion of Islamic faith is inadequate to shield even the firm believer against the knowledge of what rapid population growth is doing to the resource base. An urbanizing Sinai may seem logical to the Hadhars from the Nile Valley, but to the free-moving Bedouin it is a new and uncomfortable phenomenon.

From the Chalcolithic people of 7,000 years ago to the modern Bedouin at the close of the twentieth century, mankind's ability to live in the harsh environments of Sinai has depended upon exacting cultural adaptations that optimized existing resources and frequently tapped resources from beyond a penurious homeland. Some, like the early Egyptian, however, made little attempt to mediate the harsh environment and evolved a purely extractive symbiosis. Expeditionary forces exploited copper and turquoise, but did little to create permanent settlements. Others like Moses and the Host of Israel were only passing through, albeit on a 40-year passage.

The Nabataeans, ca. B.C. 312–32, were the first culture to exhibit nationalism in the geographic area of Sinai. Their success was based upon developing and controlling "hidden" water resources to sustain the nomadic herding of sheep and camels, cultivate a few cereals and vegetables, and practice the caravan transport of luxury products such as spices, frankincense, and bitumen across desert wastes for a considerable profit. All this was augmented by a degree of freebooting.

The modern Bedouin attempt to control their meager economic

base through a system that is still derived from a bit of nomadic herding, remote rock-walled orchards, perhaps a bit of smuggling, but more and more a dependence upon wage labor. This complex economic subsistence is facilitated by Peugeot sedans and Japanese minitrucks, but a surprising number keep camels. Perhaps they do not fully trust the Peugeot any more than they trust the urban economic systems. The human ecology of Sinai, as always, is evolving around an economy of opportunism in which the participants are never secure enough to join fully but must always retain a vestige of the past in their claim to grazing lands and oasis agriculture. Even while the heads of households live and work in urban centers or extractive industries, wives or other family members are apt to be maintaining small herds of livestock or tiny agricultural plots so as to always have a fallback position in case the wage labor falls through.

In this final chapter we will look at the prospects for a sustainable future in a land with a very meager resource base. May there always be lamb for those whose future lies in Sinai.

Between 1960 and 1990 Egypt's population increased from 26 million to 55 million, more than doubling in three decades. During the same period Sinai grew from 49,769 to an estimated 250,000, a fivefold increase. This rate is well above the maximum reproductive potential of a native population and is indicative of heavy immigration.

While there are no strictures against population control in Islam per se, there is a strong pronatal sentiment among many of its teachers. There is also a strong division of labor within the Arab culture, which views the woman's place as in the home, where her role is the bearing and nurturing of children. Polygamy is also permitted under Islamic law; however, the economic burden of additional wives may limit the full effect of this inherent growth potential.

In recent decades an increased reliability of food supplies and access to medical care have combined with existing pronatal tendencies of this society in causing rapid and frightening population growth. For Sinai the pressure is extreme because the pronatalism has been combined with the geopolitical notion that filling Sinai with more Arabs will secure it against future takeovers. This is compounded by the general misperception in Egypt that Sinai is an empty land just waiting to receive surplus population (Figure 7-1). In general, both perceptions are wrong, and Sinai must reach for the future with a great monkey on its back.

On the positive side, I believe that real economic progress has been made since the Camp David Accords. The labor market for

Figure 7-1. Africa looks at Asia, but the Egyptians on the whole view Sinai as an empty land ready to receive settlers. Point 6 Ferry departing from Ismailia—Sinai is on the opposite bank. Ferries are free for all users. No bridges have been rebuilt, and only one of the proposed highway tunnels has been built under the Suez Canal, but the land still beckons.

the Bedouin, stimulated by Israeli occupation, has continued to expand under Egyptian rule. The utilization of petroleum and natural gas fields has been an economic boon for the entire country, but Sinai's labor force especially benefits.

In 1980 the mining area at Um Bogma was reopened. Here deposits containing iron, manganese, copper, gold, lead, tin, zinc, and titanium are being utilized. In other areas of Sinai, Egypt is attempting to exploit deposits of kaolin clays, gypsum, and glass sand. The production of Portland cement and the quarrying of building stone are also increasing. One building stone in particular, alabaster (fibrous calcite), enjoys a strong foreign market for use in luxury hotels and office buildings. It occurs in six-foot-thick beds at Gebel Abu Alaga, where it is easily extracted. Sinai is rich in extractive resources, but care must be taken that their exploitation follows an orderly pattern so as to minimize the boom-bust pattern so often accompanying such development.

The agriculture of Sinai is also beginning to take on a new look. To be sure, the cropping of barley and sweet melons continues to be found wherever the Bedouin can scratch a furrow in the soil and winter rains provide enough moisture to carry the crop through to maturity. But the new technology is there also. The large-leaf Siwa alfalfa, with salt tolerance up to 5,000 ppm in the soil moisture, is helping to reclaim lands that were impossible to use before. Mangos, beloved fruit of most Egyptians, are now grafted onto sturdier disease-resistant olive stocks. The intertillage of tree crops like olives and figs is indicative of good management. Agricultural technology is well and thriving. Professors and graduate researchers are working hard to develop the best seedstocks, improve fertilizer practices, and introduce drip irrigation techniques to the Bedouin and Hadhars.

The World Food Program

The United Nations World Food Program is currently working with Sinai to improve the conditions of the rural food producers. Several interesting programs are already underway. The programs' 1989 progress reports are indicative of the directions that change should be going. They are working to improve the means of livelihood open to the Bedouin, increasing the diversification of their agriculture, improving the social and economic base, and helping people learn to benefit from education and health services.

The World Food Program–Committee on Food Aid (WFP-CFA) projects are built around four concepts: (1) land reclamation, (2) fodder shrubs and range development, (3) fruit tree plantings, and (4) arresting environmental degradation. The immediate goals of all these have focused on small plot developments, both irrigated and rain-fed. The primary objective is to encourage a shift from low-value to high-value crops—essentially from barley to fresh vegetables and tree crops. Near El Arish some 310 feddans (acres) of vegetables and 931 feddans of fruit have been planted under this program. This area is irrigated from shallow wells that serve about 4 feddans each.

Fruit tree plantations in rain-fed areas now cover 4,000 feddans, with seventy trees per feddan replacing barley and melon cropping. The WFP-CFA program subsidizes farmers in the orchard programs for 4 years while the trees are reaching fruit-bearing age. They are also building cisterns in other areas to store winter precipitation with the anticipation that a cistern can supply the needs of 2 feddans of trees each. It is planned to cover 1,600 feddans with this type of water supply.

At present there are 7,000 feddans of newly reclaimed land using Nile water that is piped under the Suez Canal through the Ahmed Hamdi Tunnel just north of the Port of Suez. This area will be divided among 1,400 settlers who will be subsidized for 4 years to allow them to become established. Additional reclamation, using Nile water, will be discussed later in sections about the Salaam Project and the extension of the Ismailia Fresh Water Canal.

Pasture and forage development is now being applied to 5,000 feddans in rain-fed areas between El Arish and Rafah that are not suitable for cropping. Two smaller demonstration areas of 500 feddans apiece are in place at El Magduba on the Wadi el Arish and at Wadi Kharan west of El Arish. In these projects 20 percent of the area is planted to fodder shrubs, with four hundred seedlings per feddan.

The three most common fodder shrubs in use are *Kochia, Haloxylon,* and *Atriplex.* These genera all belong to the *Chenopodiaceae* (saltbush or goosefoot family). Several species of the genus *Kochia* are used. The most widely planted species is *Kochia prostrata,* sometimes called summer cypress or forage kochia. The *Kochia* species are native to Eurasia, but apparently not to Sinai or Egypt, where they have been introduced for range improvement. *Haloxylon* species, as the name implies, are salt tolerant and quite useful in saline soils. *Atriplex* is probably the most widely used genus. Some species are native to Sinai. Other species have been imported from Australia. In North America, *Atriplex conescens,* commonly known as chamiso or four-wing saltbush, was used by pre-Columbian Indians for food. All three genera provide better nutrition for grazing animals than most grass species, even under the unfavorable climate of Sinai.

All grazing is prohibited for 3 years on lands planted to fodder shrubs. This should allow the new plants to become established, as well as allowing for the regeneration of natural vegetative cover. Nine plant nurseries have already been established to produce the fodder shrubs and fruit tree seedlings needed for these various projects.

Rotation of legumes with cereals is planned for 2,000 feddans. This is essentially mixing alfalfa or clover into the sequence of barley cropping. Sinai policy for newly reclaimed irrigated lands is to grow alfalfa for the first 7 years, followed by cantaloupe, cucumbers, and other vegetables needed for the urban markets.

The WFP-CFA program is also concerned with training people. At present five permanent centers are under construction at El Arish, Tor, Ismailia, Port Said, and Suez. Three mobile units will work out of Rafah, Qa, and central Sinai. These centers will train farmers in carpentry, plumbing, electrical installation, brick making, painting, and steel working. Obviously some of this will go beyond the immediate needs of agriculturalists, but it is part of the attempt to increase the sedentarization of the Bedouin, in which home construction will be an important component.

For women the training will include electrical installation, plumbing, and painting. When fully fledged, the program anticipates the training of 2,985 young people each year. Four-month courses in agricultural techniques will be given at Rafah and Qa for 390 farmers at a time. The womenfolk will be trained in embroidery and local handicrafts at a rate of 250 students each year. Within the WFP plan each Bedouin family involved is expected to construct its own house, a standard two rooms with a courtyard plus cooking and sanitary facilities.

El Arish already has a museum and demonstration project for Bedouin arts and crafts. A variety of weavings, especially camel saddle bags, are available for sale to tourists. Relevant to this is an experience by an American friend who had long lived in Egypt and who collected complete camel saddles. At a small Bedouin market he found a used camel saddle. After prolonged negotiations he purchased the saddle. An old Bedouin woman who had watched the whole process followed him out of the market. Fearing that he had in some way offended her, he finally asked what she wanted. Her reply was that she wanted to see his camel. In spite of a dearth of camels in American and European corrals, there is a reasonable market for saddle furniture and a need to keep weaving techniques alive.

It is anticipated that these programs will reach 3,000 Bedouin agriculturalists, 1,400 Hadhar settlers in the Nile water reclamation scheme east of Suez, 70 workers in the plant nurseries, and 14,500 trainees over its duration. Obviously these are small numbers compared to all the agriculturalists in Sinai, but it is a positive step from which to expand. The extension of technologies and application methods developed by the WFP should help smooth the way for Sinai's large irrigation-reclamation projects now getting underway.

Proposed Reclamation

For more than a decade, through most of the 1980's, the Nile River Basin has suffered drought. The Nile supplies over 86 percent of the 158 billion gallons of water presently used by Egypt's 55 million people each year. In the summer of 1988 the river dropped to its lowest yield since 1913. For over a century there has been a general downward trend in stream flow levels. The average yield of the river at Lake Nasser has been pegged at 92.87 billion cubic meters per year. At no time during the 1980's did it reach that level, forcing Egypt to draw on her long-term storage behind the Aswan High Dam. This caused Lake Nasser to drop 492 feet, just 7 feet above the cutoff levels where electrical generating units would have to have been shut down. The 1988 yield was only 68.5 billion cubic meters. Fortunately 1990 was a wetter year, so that water behind the Aswan High Dam rose and power generation was maintained.

Of greater concern is the long-term decline since 1870. With Egypt's population projected to reach 85 million by 2010 and plans to grow more food depending upon agricultural expansion using Nile water, the prospects are bleak. The picture is further clouded by population growth and water demand in neighboring nations of the Nile Basin—Ethiopia, Burundi, Uganda, Tanzania, Zaire, Rwanda, Kenya, and Sudan. The population of the Nile Basin is expected to increase by 100 million over the next 10 years, adding increased pressure upon the water supply.

The immediate effect of declining Nile supplies on Sinai will be small, but it is apt to have serious repercussions for land reclamation projects now under construction. A 1990 Egyptian news release, made before the Gulf War, more or less summarized the progress to that point. It stated that Kuwait and Saudi Arabia had agreed to finance the reclamation of 400,000 feddans in Sinai using water from the Nile (Figure 7-2). These loans are probably the same as those previously agreed to through the World Bank. The major portion of lands scheduled for irrigation reclamation would be served by the Salaam Canal project. The remainder would be irrigated through an extension of the Ismailia Fresh Water Canal and groundwater supplies. The Salaam Project would begin at kilometer 219 of the Damietta Branch of the Nile, where sweet water would be conveyed eastward and mixed with brackish water from the drains of existing irrigation projects. The resulting water would still be of a quality suitable for irrigation. This canal would be used to irrigate some 200,000 feddans west of the Suez Canal, but most of the water would be conveyed eastward to serve 400,000 feddans in Sinai.

The water would pass under the Suez Canal through a pressure pipe (inverted siphon) 38 km south of Port Said. The length of this pressure pipe will be 1,300 m. It will be 11 m deep under the Suez. In Sinai, the Salaam Canal will be used to irrigate 135,000 feddans in the Tenna Plain and 250,000 feddans along the north coastal area. Unfortunately the areas under consideration tend to have poor soils (compare Figure 7-2 with Figure 5-2). Sinai has no really good soils, and the fair-good soils tend to have accessibility limitations.

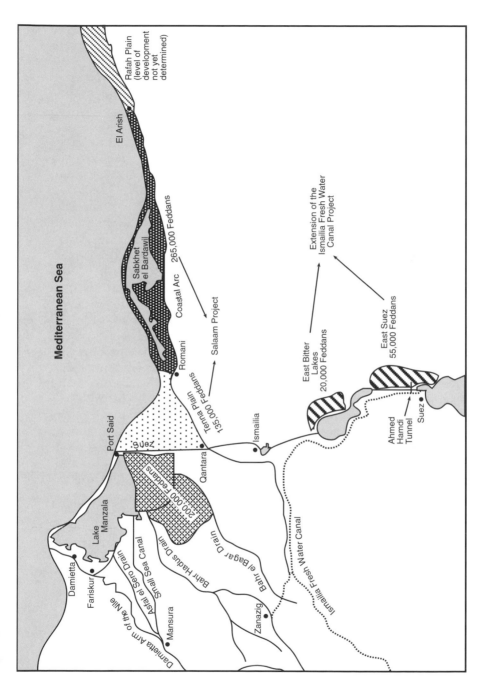

Figure 7-2. The Salaam Project and the extension of the Ismailia Fresh Water Canal Project.

It would require more pumping to lift waters high enough to serve them. Present policy is to keep the canal at elevations below 5 m in order to minimize pumping costs.

Soils of the Tenna Plain are rated poor and are mostly coarse-grained alluvium, but with some areas subject to waterlogging following heavy rains. Salinity and drainage will pose serious problems for management.

The north coastal area is divided into three units: the South Bardawil Basin, the Lower El Arish Basin, and the Rafah Plain (Sheikh Zuwayid). The South Bardawil Basin is a relatively narrow band of discontinuous arable soils of poor agricultural quality. The lower valley of the Wadi el Arish and the contiguous coastal plain have mostly fair soils intermixed with small areas of good. They tend to be deep sandy alluvium on gently sloping terrain. East of El Arish, extending to the Israeli border, is the Rafah unit, which has soils comparable to those of El Arish, but water conveyance costs would be an increasing problem. The status of Nile water for this last area is still in question.

Over 40 percent of Sinai's inhabitants live in El Arish. The demand for fresh fruits and vegetables is very high, and most of these products are brought from the Delta and Nile Valley. Efforts are being made to produce more locally. Adjacent to the city some 2,000 feddans are currently irrigated from underground aquifers, using drip irrigation and other modern techniques, to produce vegetables, melons, olives, mangos, and other fruits. This land supplies less than half of the demands for fresh produce. The government wants El Arish to become self-sufficient in perishable food products. This goal is complicated by increasing salinity in the groundwater now being pumped for irrigation. High priorities exist to extend the Salaam Canal as quickly as possible into the coastal area.

It would be highly desirable to extend the system all the way to El Arish before applying water to the Tenna Plain, in order to take advantage of the somewhat better soils found there and supply the known local demand for vegetables and fruits.

It should be noted that El Arish has had a problem with the use of brackish water for domestic consumption. In April 1988 a pipeline conveying fresh water from the Nile reached El Arish, greatly relieving the serious problem there. The city now had dual distribution systems for all domestic water. Each household has a single spigot for the sweet water. Toilets, baths, and all other uses are served by the brackish water distribution system. One evening when I was first in El Arish, I ran out of bottled water for drinking purposes. So I ran a bottle of water from the bathroom faucet, treated it with trusty iodine, let it sit for 8 minutes, and took a big swig. It nearly made me ill. The salt content was unacceptably high. I learned the hard way about the dual water conveyance systems of El Arish.

It should be noted that the Bedouin can obtain drinking supplies at various places along the pipeline without charge. This is already changing the demographic characteristics of the Mediterranean Littoral. By contrast, the Arisha (citizens of El Arish) pay a premium for sweet water in the city.

East of the Suez Canal, in the middle and southern sectors, it is planned to deliver Nile water through an extension of the Ismailia Fresh Water Canal (Figure 7-2). This unit will ultimately serve 30,000 feddans east of the Great Bitter Lake and 55,000 feddans east of Suez. Unfortunately soils in these areas are also classed as poor, so production is apt to be marginal. Vegetables from these units will probably move westward to meet demands in Ismailia and Suez. Some water is already moving by pipeline through the Ahmed Hamdi Tunnel just north of the city of Suez. The project is so recent, however, that there is no indication of its productivity.

Other projects under consideration will be dependent upon groundwater resources. The largest of these is the Qa Plain proposal, where 2,000 to 4,000 feddans are slated for development. Most production here will be for the local needs of nearby oil fields, mines, and tourist areas. Beyond that, fruits and vegetables would move toward the Suez market. The Egyptians are quick to acknowledge that even if groundwater is adequate, the area will suffer from high transport costs and the inability to attract farmers. Egyptians are gregarious people, and it will be difficult to persuade Fellahin to live in this remote area. One advantage of the Qa is in slightly better soils that will make it more productive than comparable developments in the Tenna Plain or the Bitter Lakes Basin.

To utilize these reclaimed lands the government has decided upon three patterns of ownership: small holdings with 5 feddans per family unit, to be used in vegetable production, graduate holdings with 10 feddans per unit and larger capital investment, and management-intensive units for specialized industrial crops. Five feddans is considered optimal for a family unit who control their own water supplies and practice a labor-intensive vegetable production. These units will mostly use drip irrigation. The 10-feddan allotment to graduates of agricultural schools is an effort to encourage technological training. The larger capital- and management-intensive units are to be used primarily with industrial crops such as cotton. The numbers of farms in the various size units have not been determined. The government hopes to encourage a mix of small and large plots so that small holders might provide a labor source for the larger units and small holders gain access to supplemental income.

For the Bitter Lakes unit of 30,000 feddans, it is anticipated to have a headquarters town, plus five service-market villages and up to twenty-five satellite villages that will accommodate two hundred smallholder families each. Initially 5 percent of the area will be devoted to the 5-feddan units, but as it develops that percentage will increase to 16. Past experience in the Delta and around El Arish has proven such a pattern to be quite feasible.

In all projects amortization by individual farmers will be spread over a 25-year period during which no charges are made for water. During the first 3 years, small farmers will make no payments. Assuming farm yields at 75 percent of those obtained in the northeast Delta, it is projected that returns in the first year will be 525 LE (Egyptian Pounds), increasing to 617 in the tenth year and 1,037 at the termination of the land payments. This is thought to be sufficient to assure that most farmers would continue on the project. It is not known if water costs will be assessed once the land is paid for. For us in the industrial world the returns seem meager in the extreme, but for dispossessed rural workers it may be a welcome opportunity to obtain their own bit of land.

Fishing

Fishing, like agriculture, has room for improvement. Dr. Samir Ghoneim, Head of Fish Research as well as Dean of the Faculty of Environmental and Agricultural Sciences, Suez Canal University, continually pushes for the best techniques of fish breeding, habitat maintenance, and pollution control. Sabkhet el Bardawil, the main fishing grounds of north Sinai, was once completely land-locked, with problems of increasing salinity and declining fish populations. Since 1955 two channels have been opened to the Mediterranean to allow better circulation. This has stabilized the levels of salinity and improved nutrient flow. From Bardawil in the north to the coastal waters extending from Tor to Nuweiba in the south, conditions are constantly monitored (see Table 7-1). Bedouin fishermen are taught how to fish so as to maintain stocks. At the same time large ice and freezing plants have been constructed at Tor and El Arish to help preserve the catch, to minimize waste, and allow optimal market utilization. While in Sinai we had many tasty meals of Bardawil bream (Figure 7-3).

Table 7-1. *Sinai Fisheries Data*

Fishing Area	Catch (tons)	Type of Fish	%
Gulf of Suez, Gulf of Aqaba, and Red Sea	20,000	Sardine	30
		Sepia	18
		Sea bream	15
		Red mullet	12
		Lizard fish	10
		Shrimp	5
		Misc.	10
Sabkhet el Bardawil	2,500	*	

*Breakdown by type is not available for Sabkhet el Bardawil, but the dominant species include sea bream, sepia, and red mullet.

Sources: El Feky, pp. 51–57. Ettewa, pp. 164–171.

Note: Catches in both the Red Sea area and Sabkhet el Bardawil have remained rather constant since 1980.

Advancing with Sedentarization

Since the return of Sinai, Egypt has pushed for the sedentarization of the Bedouin. Programs of housing aid have been the most visible elements of these policies. Stories of Bedouin utilization of government housing are reminiscent of U.S. programs for the Navajos. It seems that cement block and similar construction were not happily received. Thermal transmission characteristics made these structures hot in summer and cold in winter. The Bedouin would frequently use the house for storage and build an arbor with attached courtyard with palm fronds for privacy but still capable of taking advantage of any air movement and nocturnal cooling to mitigate summer heat. In winter the low ceilings of the goat hair tent helped hold the warmth better than the conventional, poorly insulated structures.

The increasing availability of wage labor has done more to settle the Bedouin in permanent locations than all the governmental policies combined. Drought and cold winters, especially in the high country over the last few decades, have served to reduce herd sizes and made it logical for the nomads to reduce the extent of their transhumance. In some instances they were able to operate from a fixed base.

There are advantages to permanent housing, but the Bedouin have been slow to accept them. With proper architectural design, permanent dwellings can be made far more comfortable, winter or

Figure 7-3. Arab fishermen at the Sabkhet el Bardawil.

summer. These advantages will be discussed below. Meanwhile the Bedouin are faced with making both a cultural and economic adjustment to declining herd sizes and the acceptance of permanent domiciles. Two things are helping, the aforementioned wage labor and the availability of motorized transport, especially Toyota trucks. It is still startling to see a small truck with a full-grown camel kneeling and strapped in the bed on its way to market. The destination of the camel is another interesting question. It may be on the way to becoming a riding, transport, or draft animal for another Bedouin, but just as likely is the possibility of its going to an Egyptian meat market. At present Egypt consumes about fifty thousand camels per year. Many of the animals come from Nubia and Sudan, but Sinai is also a supplier. The future economics of camel beef should not be overlooked in the changing structure of economics. Cost-effectiveness studies need to be made on the efficiency of camel production for meat.

Housing to Put Climate to Work

The increasing role of permanent housing in Sinai has the potential to combine traditional values with simple futuristic energy conversion and waste disposal technologies. Research and development by people like Ken Kern have amalgamated concepts that could work well in Sinai's environment. A combination of passive solar heating and undergrounding or earth berming can be used to optimize the intense radiation (Tables 4-1 and 4-2) found here. Figure 7-4 shows solar concepts applied to a simple structure that would be well suited to Sinai's programs for settling the Bedouin. Styling could follow the lead provided by the Tourist Village at St. Catherine, where there is an attempt to make the bungalows resemble Bedouin tents.

Solar collector panels can provide for water heating and augmentation of passive heating through south-facing windows and Trombe walls. The bottom floor of the south-facing room would be designed with maximum insulation and would be largely closed off for winter living. The north-facing room would be situated so as to maximize ventilation and would be the center for summer activity.

Small parabolic solar cookers (skier stoves) could reduce fuel demands. Such units can easily boil water, make soup, and generally perform any cooking operations done on an open flame. Since most Bedouin cooking is done over the flames rather than in ovens, adaptation should pose few problems. Winter cooking could be done inside. In the heat of summer, simply move outside.

The integration of solar heating and parabolic cookers combines the best ways to insure winter warmth and cooking fuel while reducing the present practice of gathering any available biomass for fuel. Thus solar heat is "less than fuel wood" at the same time it can be "more than fuel wood" because it saves precious organic matter, which can be returned to enrich the soil. It will take time, but it is one of the great environmental needs of Sinai.

Small-scale electrical power generation could be based on wind chargers and augmented with photovoltaic cells. Enough power for a compact refrigerator, radio or television, and minimal lighting could be generated in this way. With the heavy costs of building power grids and large-scale generating units, this is a logical alternative for scattered villages and isolated homes. In some cases it might be more cost-effective to build larger wind generators for a village, with some type of backup system to cover the periods of calm.

Waste Conservation

Human wastes may be dangerous or beneficial depending upon disposal systems. For rural Sinai this is generally no problem except around watering spots where individuals are not always considerate of those who come later. In the cities, public health codes and enforcement have reduced water-borne diseases to a minimum.

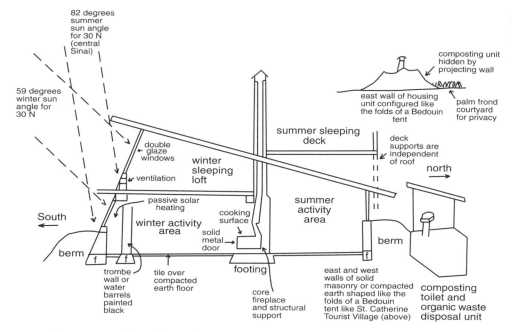

Figure 7-4. A basic house for Sinai, two rooms, sleeping loft, deck, and outside composting toilet (adapted from Kern).

The construction of permanent residences in rural areas has the potential to increase human waste problems or turn those wastes into valuable products. Dry composting units, sometimes designated Clivis systems, are well suited for housing units in rural and village areas. Composting not only kills pathogenic organisms, it also produces a rich organic humus to augment the low-humic soils.

Urban sewage and garbage, because of their volume, are not as easily dealt with. There is some truth in the old saw that "dilution is the solution to pollution." By contrast, concentration complicates reclamation. Fortunately the nature of waste in less-developed countries makes it easier to utilize. In El Arish the urban household waste is high in organic and putrescible matter. High moisture content and density make the waste suitable for composting to produce methane gas, fuel pellets, and soil conditioners. Some composted materials can be used for fish feed in nearby Sabkhet el Bardawil. In some countries hogs are also incorporated into the conversion system. Unfortunately Islam rejects hogs as unclean meat, so they would be unfeasible here. One logical alternative would be the use of fowl or water birds instead, so that food production could still be introduced at this state of the resource cascade.

Methane production could be used in conjunction with El Arish's solar generating units to even out the flow of electrical energy. It could also be used to desalinate the brackish groundwater, which requires less energy than to remove salt from sea water. Thus there are numerous ways to integrate waste reclamation into the resource flows of Sinai.

Education

Egypt has always led the Arab world in education. Literacy is a high 44 percent, and Egyptian teachers are hired throughout the Middle East to conduct primary and secondary education. Egyptian universities train people from all over the world, but especially those from north Africa and southwest Asia. Education is already a major Egyptian export and one that can continue to grow and improve.

Since the accords of 1979, Egypt has expanded education services to Sinai. The Suez Canal University at Ismailia has the responsibility for filling Sinai's needs for higher education. The main campus at Ismailia has some ten faculties (colleges) covering a wide range of disciplines from agriculture through education and medicine. At Suez and Port Said are technical faculties that service the specialized needs of those busy ports.

Since Israel's withdrawal from Sinai, the university has established a Faculty of Education and the new Faculty of Environmental Agricultural Sciences at El Arish. Research stations at St. Catherine and Sharm el Sheikh complement the research of various faculties throughout the university and provide specialized focus on mountain environments, native resource systems, oceanic systems, and fish resources.

The Suez Canal University, with its highly qualified faculty, is in an excellent position to train students from oil-rich Arab neighbors. Such is a complementarity that would bring income to Sinai and expand education to people in places like the United Arab Emirates, Saudi Arabia, and even Yemen. The emphasis should continue to be on science and technology, particularly in ways that can help the Middle East. Soil science, agriculture, biological sciences, water resource technology, geology, marine resources, industrial arts, and geographic information systems are the types of offerings that should be readily available. At the same time it would be good to have courses in Arab culture, archaeology, and even specialized offerings in ancient manuscripts and Orientalism. Sinai has much to offer in these last three areas.

Capturing the Professionals

By way of constructive criticism, I would like to mention a problem that hurts Egypt and Sinai, both in higher education and most other professional areas. As in many less-developed countries there is a tendency to concentrate the best businesses, the exciting people, the apex of the arts, and much else in the central capital city. Cairo is an exciting, vibrant city. Regional centers suffer in contrast to the great capital, and most qualified people want to live at the center. Many professors employed in Ismailia, including department heads at Suez Canal University, live in Cairo. They put in a minimum of time at the university and tend to focus their lives on Cairo. At El Arish there is a step-down of this model. Professors at the new faculties want to live in Ismailia and come to El Arish for a minimum of time. To be sure, there is presently not enough housing, but the governorate is striving to meet this need. Yet the problem persists. The wives of many government officials, including the wife of the Governor of North Sinai, live in Cairo. In a free society it is difficult to dictate to these people where they must live, but some way needs to be found to capture the professionals in the setting where they earn their money and contribute their expertise. It is like the Bedouin problem of "home of the heart." For these professionals there must be found a way to transfer their hearts to El Arish, Tor, or wherever their services are rendered.

Tourism

El Arish is a great beach city. Someday it may well compete with Alexandria for Egypt's vacationing masses who want to cool off in the Mediterranean, do a bit of fishing, or lie on the beach and catch a few rays. At present there are far fewer people competing for its relaxing shores. I for one would rather relax on the beaches at El Arish than those of Alexandria. For the shell connoisseur it can be equally exciting. The violet sea snail (*Janthina*), the nicely sculpted *Murex* species, the vivid sunray (*Macrocallista nimbosa*), and the moon shells (*Polinices*) only hint at the variety to be seen.

On the beach is the luxurious, five-star Oberoi Hotel with full Western dining facilities. Within a block or two from the beach one can dine on fish specialties or traditional Arab fare of ful, round bread, and tahina. El Arish is a tourist spot on the verge of discovery.

South of El Arish on the way to St. Catherine there are several places worthy of a tourist's attention. The lush Oasis of Feiran, with its ancient monastery, its date and deciduous fruits, and access to rugged Gebel Serbal, deserves more attention than it now receives. The crown jewel for Sinai tourism is, of course, the great fortified convent of St. Catherine. Airport, tourist village, souvenir shops, and friendly people make the Western traveler feel right at home, yet the culture and exotic landscapes provide a class experience for those who make an effort to get away from the tourist traps. Unfortunately this area is in real danger of being loved to death. The tourist village already has 250 rooms. Just 6 miles north a billboard announces the construction of 500 villas. All that is lacking is a cable car to the top of Mount Sinai (Gebel Musa) to complete a real circus in the "Great and Terrible Wilderness." Little help that powerful rulers from Justinian to Golda Meir and Hosni Mubarak have taken it under their wings. The seventeen monks at St. Catherine are being overwhelmed by 1,500 visitors each day— that is, 565,000 per year. Sinai has much to offer, but Sinai has much to lose.

At the tip of the peninsula is reputedly the best scuba and snorkeling in the world. For one who has dived at Bora Bora, Moorea, Guaymas, the Great Barrier Reef, and in the murky kelp beds off San Diego, it is truly a wonder of marine life and beauty. From Ras Mohammed to Sharm el Sheikh limpid blue waters, colored coral reefs, and rare fish excite the wonder of terrestrial man. The waters off Ras Mohammed are now protected by a Marine National Park. We must hope that the park will preserve this great environment from undue encroachment.

Fully equipped diving centers, dive boats, hotels, and youth hostels are geared to Western tourism. Farther up the Gulf of Aqaba side, Dahab and Nuweiba offer diving, plus some of the world's best wind surfing set against a backdrop of rugged mountains that run right down to the sea except where isolated shelf-beaches give access to the water (Figure 7-5). The view eastward across the narrow gulf is equally beautiful where the blue waters touch the purple mountains of the Hejaz of Arabia. If these things are not enough, there is access to the Red Sea Islands. More than twenty-four islands, including Zabargad, Abu Minkar, and Shedwan, offer the truly out of the ordinary escape.

Of the Future

Sinai is a great land, but it is a fragile land which must be accepted on its own terms. Now at the end of the twentieth century, it faces

Figure 7-5. Wind surfing on the Gulf of Aqaba near Nuweiba.

great danger. Circumstances are changing so rapidly that there is little time for the land or the people to adjust to the new stresses. Water resources, always crucial, are under new pressures as Egypt seeks to feed its explosive population growth by rapid development of irrigation reclamation, especially in Sinai. Plans to divert more water from the Nile for use in Sinai are apt to be hampered by declining flows and more competition for those waters by other nations within the greater Nile Basin.

Even as these pressures are building, the population of Sinai is expanding at two-and-a-half times the rate for Egypt as a whole, largely because of governmental policies for population relocation. To some extent the Bedouin of Sinai have benefited from economic growth, more jobs, and more markets for their products. Tourism is growing, but the very qualities that attract visitors are being destroyed by the economic and political pressures now tearing at Sinai's environment. In the long run it is doubtful that these desert habitats can withstand the pressures now being exerted against them. Only exacting planning and implementation can solve the short-range problems. In the longer picture, population control and careful utilization of resources can retard the constant erosion of environmental quality in this exotic land.

Bibliography

Atlas of World Agriculture, Vol. IV (Africa). 1969. Navara, Italy: Instituto Geografico de Agustine.

Bailey, Clinton. 1984. "Bedouin Place-Names in Sinai," *Palestine Exploration Quarterly*, January–June, 42–57.

Barron, T. 1907. *The Topography and Geology of the Peninsula of Sinai (Western Portion)*. Cairo: National Printing Department.

Bartlett, S. C. 1889. *From Egypt to Palestine, through Sinai, the Wilderness and the South Country*. New York: Harper and Brothers.

Bartov, Y., et al. 1985. "The Geological Structure of Sinai," *Atlas of Israel*, 3d ed. Tel-Aviv: Survey of Israel, and New York: Macmillan Publishing Co.

Beadnell, H. J. Llewellen. 1927. *The Wilderness of Sinai*. London: Edward Arnold and Co.

Besant, Walter. 1883. *The Life and Achievements of Edward Henry Palmer*. London: J. Murray.

Burchfield, B. Clark. 1983. "The Continental Crust," *Scientific American* 249, no. 3 (September): 130–142.

Burckhardt, J. L. 1822. *Travels in Syria and the Holy Land*. London: John Murray.

Chesnoff, Richard Z. 1988. "When Water Feeds Flames, Growing Shortages in Mideast Add to Regional Tensions," *U.S. News and World Report*, November 21, 47–48.

Dames & Moore's Center for International Development and Technology. 1982. *Sinai Development Study, Phase I, Final Draft Report*, Vol. III (Agriculture and Fisheries), Submitted to the Advisory Committee for Reconstruction, Ministry of Development, Arab Republic of Egypt, US-AID Grant No. 263-0113.

Dan, J., et al. 1982. "Evolution of Reg Soils in Southern Israel and Sinai," *Geoderma* 28, nos. 3, 4.

Danin, Avinoam. 1979. "The Flora: Wild and Cultivated Plants," in Beno Rothenberg (ed.), *Sinai: Pharaohs, Miners, Pilgrims and Soldiers*. New York: Joseph Binns.

———. 1983. *Desert Vegetation of Israel and Sinai*. Jerusalem: Cana Publishing House.

El Feky, A. 1986. "A Review of Egyptian Mediterranean Fisheries," *Report of the Technical Consultation on Stock Assessment in the Eastern Mediterranean.* Rome: FAO Fisheries Report 361.

Engler, Adolf. 1924. *Syllabus der Pflanzenfamilien,* Auf 9 und 10. Berlin: Gebruder Borntraeger.

Ettewa, I. 1988. "Report on the Egyptian Fisheries Focusing on the Mediterranean," *Report of the Second Technical Consultation on Stock Assessment in the Eastern Mediterranean.* Athens: FAO Fisheries Report 412.

Eyal, M. 1985. "Geological History of Israel and Sinai," *Atlas of Israel,* 3d ed. Tel-Aviv: Survey of Israel, and New York: Macmillan Publishing Co.

FAO/UNESCO. 1968. *Definitions of Soil Units for the Soil Map of the World.* World Soil Resources Report 33. Rome: World Soil Resources Office, Land and Water Development Division, FAO.

———. 1975. *Soil Map of the World, 1:5,000,000,* Vol. VII, Africa. Paris.

———. 1977. *Soil Map of the World, 1:5,000,000,* Vol. VII, South Asia. Paris.

Flint, R. F. 1973. *The Earth and Its History.* New York: Norton and Company.

Furley, Peter A., et al. 1983. *Geography of the Biosphere: An Introduction to the Nature, Distribution and Evolution of the World's Life Zones.* London: Buttersworth.

"Geology of Sinai, Sheet 10." 1985. *Atlas of Israel,* 3d ed. Tel-Aviv: Survey of Israel, and New York: Macmillan Publishing Co.

Good, Ronald. 1947. *The Geography of the Flowering Plants.* London: Longmans, Green and Co.

Goodman, Steven M., and Peter L. Meininger (eds.). 1989. *The Birds of Egypt.* Oxford: Oxford University Press.

Greuter, W., et al. (eds.). 1988. *International Code of Botanical Nomenclature.* Konigstein, Germany: Koeltz Scientific Books.

Griffiths, J. F., and Kit Soliman. 1972. "The Climate of the United Arab Republic," *World Survey of Climatology,* Vol. 10 (Climates of Africa). New York: Elsevier.

Harland, W. Brian, et al. 1989. *A Geologic Time Scale 1989.* New York: Cambridge University Press.

Hart, Henry Chichester. 1893. *Flora and Fauna of Sinai, Petra and Wady 'Arabah.* London: Palestine Exploration Fund.

Hobbs, Joseph J. 1989. *Bedouin Life in the Egyptian Wilderness.* Austin: University of Texas Press.

Holmes, Sandra. 1983. *Outline of Plant Classification.* London: Longman.

Hume, W. F. 1906. *The Topography and Ecology of the Peninsula of Sinai.* Cairo: National Printing Department.

Hutchinson, J. 1959. *The Families of Flowering Plants* (Vol. I, Dicotyledons). Oxford: Clarendon Press.

Irving, Thomas B. (trans.). 1985. *The Qur'an, The First American Version.* Brattleboro, Vt.: Amana Books.

Katz, Donald. 1979. "The Wilderness of the Howling Zeal," *GEO* 1 (December):6–34.

Kern, Ken. 1975. *The Owner Built Home.* New York: Charles Scribner's Sons.

Köppen, W. 1923. *Die Klimate der Erde; Gundriss der Klimakunde.* Berlin: Walter de Gruyter Co.

Kummel, Bernhard. 1970. *History of the Earth.* San Francisco: W. H. Freeman and Co.

McArthur, Durant. 1991. Personal interview regarding forage shrubs used in Sinai, October 9. Shrub Science Laboratory, Provo, Utah.

Mathews, Samuel. 1967. "Science Explores the Monsoon Sea," *National Geographic Magazine,* October, 554–575.

Meissner, Rolf. 1986. *The Continental Crust: A Geophysical Approach.* San Diego: Academic Press.

Moffett, George D., III. 1990. "Middle East's Cup Runneth Dry," *Christian Science Monitor,* March 8.

Moon, F. W., and H. Sadek. 1921. *Topography and Geology of Northern Sinai,* Pt. 1, Session 1919–1920. Ministry of Finance, Egypt, Petroleum Research Bulletin No. 10. Cairo: Government Press.

Morrow, Lance. 1990. "Trashing Mount Sinai," *Time,* March 19, 92.

Musil, Alois. 1928. *The Manners and Customs of the Rwala Bedouins.* New York: American Geographical Society, Oriental Explorations and Studies, No. 6.

Neev, D., et al. 1987. *Mediterranean Coasts of Israel and Sinai.* New York: Taylor & Francis.

Parker, Sybil P. (ed.). 1982. *Synopsis and Classification of Living Organisms.* New York: McGraw-Hill Book Company.

Post, George E. 1933. *Flora of Syria, Palestine and Sinai,* 2d ed. (extensively revised and enlarged by John Edward Dinsmore). Beirut: American University Press.

Rawlinson, George. 1880. *History of Ancient Egypt.* New York: American Publishers Corporation.

Ritter, Carl. [1866] 1972. *The Comparative Geography of Palestine and the Sinitic Peninsula,* Vol. 1 (translated and adapted by William L. Gage). St. Clair, Mich.: Scholarly Press.

Rudloff, Willy. 1981. *World Climates.* Stuttgart: Wissenschaftlinche Verlagsgesellschaft.

Siever, Raymond. 1983. "The Dynamic Earth," *Scientific American* 249, no. 3 (September):46–55.

Smith, A. G., et al. 1977. *Phanerozoic Paleocontinental World Maps.* Cambridge: Cambridge University Press.

Soil Conservation Service. 1973. *Soil Survey, San Diego Area, California,* Part II. Washington, D.C.: Department of Agriculture.

———. 1975. *Soil Taxonomy.* Washington, D.C.: Department of Agriculture.

———. 1985. *Keys to Soil Taxonomy.* Technical Monograph No. 6, Soil Management Support Services, prepared for the United States Department of Agriculture and Agency for International Development. Ithaca, N.Y.: Agronomy Department, Cornell University.

Tchernov, Eitan. 1979. "The Fauna, Meeting Point of Two Continents," in

Beno Rothenberg (ed.), *Sinai: Pharaohs, Miners, Pilgrims and Soldiers.* New York: Joseph Binns.

Thompson, B. W. 1965. *The Climate of Africa.* New York: Oxford University Press.

——. 1975. *Africa: The Climatic Background.* Ibadan, Nigeria: Oxford University Press.

Tristram, H. B. 1884. "The Fauna and Flora of Palestine," in *The Survey of Western Palestine.* London: Palestine Exploration Fund.

UNESCO. 1963. *Bioclimatic Map of the Mediterranean Region* (East Sheet). Paris.

——. 1970. *Ecological Study of the Mediterranean Zone, Vegetation Map of the Mediterranean Zone, Explanatory Notes.* Paris.

——. 1977. World Meteorological Organization. "Meteorological Aspects of Solar Radiation as an Energy Source," *Annex–World Maps of Relative Global Radiation,* Technical Note No. 172, WHO, No. 557.

U.S. Army Map Service. 1958. "Tor," Map NH 36-15 (Series P502), 1:250,000.

Wallace, Alfred Russell. 1876. *The Geographical Distribution of Animals.* New York: Harper Brothers.

Windley, Brian F. 1977. *The Evolving Continents.* New York: John Wiley & Sons.

Zohary, M. 1935. "Die Phytogeographische Gliederung der Flora der Halbinsel Sinai," *Beihefte zum Botanischen Centralblat* 52, Kassel, Germany.

Index

Abu Aweigla, x, 7
Abu Durba, 7
Abu Kiseib, Moyet, 42
Abu Mesud, Gebel, 40
Abu Rudeis, x, 7, 35, 62
Abu Rumail, Gebel, 43, 44
Abu Shiah, Gebel, 42
Abu Tarfa, xii
Abu Zenima, x, 7
Adakkar, Gebel, 40
adaptations to optimize resources,
    114
Adh-Dhayga (see Halal Narrows)
aeolian sands, 24, 26
aeolian soils, 98
Afar Triple Juncture, 17
African Plate, 13, 16, 18
Afro-Arabian Plat, 17, 25
agricultural potential of soils, 81,
    82
Ahmed Hamdi Tunnel, 117, 121,
    123
Ain, Wadi, 35
Ain Akhdar, xi, 8
Ain Furtaga, 7
Ain Hudera, 8
Ain Sudr, 7
Ain Um Ahmed, 8
Ain Yirga, xii, 8
Ajeleh, Wadi, 42
Akhdar, Wadi, 35, 48
Akhdar Pass, 35
Aleyat, Wadi, 41, 42
Al Ismailiyah, 5, 6

al juwra (pit), 29
al-kez (the terrible summer), 68,
    112
allogenic water, 66
Amalekites, 49
Anatolian Plate, 13, 19
annual grass, most important,
    98
Aqaba, 12
  Drainage, 49
  Foreshore, 3, 10, 38
  Pull-Apart Basin, 18, 19, 50
  Wadi, 48
aquifers, 122
Araba, Gebel, 49
Arabian Peninsula, 10
Arabian Plate, 3, 13, 16–19
Arabian Sea, 56
Arabo-Nubian Massif, 12, 16
arid areas, 60–61
Aridisols, 82–86
aridity, 3, 59–61
Asfal el Sero Drain, 121
Ashraf, 9
asseif (the good summer), 68
assferi (rainy season), 68
assmak (spring), 68
assta (winter), 68
As Suways, 5, 6
Atantur, Ras, 38
Atiya Fault, 36, 37
Atlantic Subtropical High Pressure
    System, 3, 53
Ayun Musa, 7

Baba, Wadi, 33, 35
Bahr el Bagar Drain, 121
Bahr Hadus Drain, 121
Bardawil (*see* Sabkhet el)
barometric observations, 46
Barron, Thomas, 46
basic house, 128
batholithic intrusions, 12
Beadnell, H. J. L., 19, 22–24
Bedouin climatology, 68
Bedouin year, 68
biological activity, swings of, 113
birds
    accipiters, 109
    cranes, 109
    cream-colored courser, 109
    egrets, 109
    fishing hawks, 109
    ground, 107
    little bustard, 107
    little owl, 109
    migratory, 108
    osprey, 109
    ostrich, 107
    quail, 107
    raptors, 109
    raven, 109
    reef heron, 109, 110
    sand grouse, 107
    sand partridge, 107
    Tristam's grackle, 106
    tropical sunbird, 106
    water, 107, 109
    white stork, 107
Bir Gifgafa, x, 7, 8
Bir Hasana, x, 3, 7, 8
Bir Tamada, 8
Bitter Lakes unit, 121, 124
Biyar, Naqb, 35, 40
Biyar, Wadi, 35
Blue Mountain, 1
Bodhia, Gebel, 24
bone bed (Gebel Safarial), 37
botanical nomenclature, 90
brackish water, 122
bread of heaven, 94
broom, white, 98

Brow of Musa (Sufsafa, Ras), 43, 44
Bruk, Wadi, 30, 48
Burckhardt, J. L., 46
Bur Fuad, 7
Burial place of the Tables of the
    Law, 43
burrowing as adaptation, 105
Bur Said, 5, 6

Calciorthids, 72, 82
calderas, 12
Camborthids, 72, 83
camel beef, 126
camel-riding scientific explorers,
    19
camel thorn, 94
camel tick–raven symbiosis, 109
Camp David Accords, 5, 115
cancer root (*see* dodder)
capturing the professionals, 130
Carboniferous sandstone, 21
Carboniferous System, 13
cation exchange capacity, 78, 79,
    81, 85
cattail, 102
Cenomanian-Turonian limestone
    and dolomite, 20, 21, 36
Chalcolithic people, 114
chalk, 22
chamiso, 118
changing demographics, 123
charcoal, 98
cisterns, 117
climatic classification, Koppen,
    60–64
club rush, 104
coastal plains and tetonic stability,
    25
colonizing species, 102, 103
common reed, 102
continental collision, 12
continentality, 56
contraband, 8
Convent of Forty Martyrs (*see* El
    Arbain)
convergent plate boundary, 18, 19
coral reefs, 37, 101

Cretaceous materials, 24, 37
crossroads of continents, 1
crustal shortening, 19
cuesta landforms, 30
cul-de-sac-wadis, 31
cultivable soils, 74
cultural adaptations, 114
cultural regionality, 9
cyclonic movement, 59

Dahab, x, xii, 7, 8, 38, 50
Dahab Drainage, 47, 49
Dalal, Gebel, 35, 40
Dalfa, Gebel, 34
Damietta Arm of the Nile, 121
Dan, J., 86
Danin, Avinoam, 90, 95, 96, 103, 104
Darb el Haj (Pilgrim Road), 8–10
date palm, 100
Dead Sea Drainage, 27, 46, 47, 50
Dead Sea Pull-Apart Basin, 18, 19
Dead Sea Rift, 17, 18
Deir, Gebel, 43, 44
Deir, Wadi, 44
descendants of Mohammed, 9
desert climates, 61
desert limit (climatology), 61
desert locusts, 112
desert mountain climates, 61–63
Desert of the Exodus, 31
Desert of the Wandering, xii
Desyet Fureiah, 44
Dike Country, 39
dikes, 12, 17, 20–24, 38–39
dike system, longest, 24
divergent plate boundary, 18, 19
diversification of agriculture, 117
Dividing Valleys, 2, 27, 28, 31, 33, 35, 36, 40, 48
division of labor, 115
dodder, 101
dolerite intrusions, 39
domal configuration, central Sinai, 23
domal uplift, 17
domestic animals, 4

doum palm, 101
drainage units, 46, 47
drip irrigation, 117
drought in Nile Basin, 119, 120
dry compositing, 128
dryness limit (climatology), 61
dune ridges, 24
Dune Sheet, 27–30, 33
dune stabilization, 80
Durorthids, 83
dynamically induced pressure system, 53

East African Rift, 17
economic progress since Camp David, 115
ecosystem complexity, declining, 3
education, 129, 130
education as an export, 129
Egma, Gebel, 23, 31
Egma Escarpment, 23, 28, 31, 33, 35, 36, 48, 50
Egma Limestone (Lower Eocene), 23
El Arbain (also Deir el Arabian), 44
El Arish, x, 5, 7, 29, 51, 58, 67
    Basin (area), 31
    Campus, Suez Canal University, vii
    Drainage Basin, 46–48
    freshwater pipeline, 122
Elat, x, 67
Elat materials, 12
electrical conductivity (soils), 78, 80, 85
Elijah's Chapel, 45, 46
El Magduba demonstration area, 118
El Shatt, 7
endemism, 89, 91
Entisol profiles of Wadi el Arish, 73, 75, 77, 78
Entisols, 71, 73, 75–82
environmental pollution, 3
Eocene materials, 20–23, 31, 36
Eocene Nile, 19

Equatorial Trough of Low Pressure, 53, 55
erg, 71–72
erosional geomorphology, 26
Er Rahah, 95
Euramerica, 13
evapotranspiration, actual, 67
evapotranspiration, potential, 51, 59, 67
evolution of vegetation, 95
exclusive species, 102
exploratory drilling, Gebel Maghara, 22
extractive resources, 116
extractive symbiosis, 114
eyelash (*see* shok-e-dhab)

faculties (colleges), 129
Faculty of Environmental Agricultural Resources, vii, 76, 129
Fatima, 9
fault cliffs, 19
faunal regions
  effect of Tethys Sea, 105
  Ethiopian Region, 106
  Palearctic Region, 105
Feiran, 7, 8
  Drainage, 47
  materials, 12
  Oasis, 33, 35, 41, 42, 131
  Wadi, 12, 35, 37, 39, 42, 95
Feirani, Gebel, 40
Feiran-Serbal area, 42
Fellahin, 10
felsite intrusions, 39
Fera, Gebel, 43, 44
fertility management (soil), 86
fish research, 124
floodplain soils, 75
floristic regions, 88
fluviatile conglomerates, 13
forage kochia, 118
foraminifera (*see* nummulitic limestones)
four-wing saltbush, 118
French tamarisk, 94

fringing reef, 38
frogbit, 111
frogs, 111
fruit husbandry, 9
fuel, 100
Furley, Peter, 106
future of Sinai, 131, 132

gabro, 12
Garf, Wadi, 33, 35, 49
Gebel Musa Area, 43, 44
Gebel Serbal Area, 40–42
Gebel Zebir (Saint Catherine Cluster), 43
geographic nomenclature, 2, 4
geologic column of Sinai, 14, 15
geomorphic regions, 27
geopolitics, 115
geosyncline, Senonian, 22
Geraia, Wadi, 33, 34
Geziret Faraun, 38
Gharandal, Wadi, 30, 95
Ghoneim, Samir, 124
Giddi
  Drainage, 47
  Massif, 33
  Pass, x
  Wadi, 30
Gineina, Ras, 35, 40
GIS (geographic information systems), 19, 129
Gondwanaland, 13
goosefoot (*see* saltbush)
governorates, 5, 6
granodiorite, 12
grapes, 8
great crescent of loose sand, 29
Great Soils Group, 79
groundwater reliability, 49
Gulf of Aden Rift, 17
Gulf of Aqaba, 57
Gulf of Aqaba Drainage, 46
Gulf of Suez Drainage, 46, 49
Gulf of Suez Rift, 17
gum Arabic, 95
gum resins, 94, 95
Guran, Ras, 29

Habashi, Gebel, 40
habitat competition, 106
habitat maintenance, fish, 124
Hadahid, Wadi, 37
Hadhars, 114, 117
Hadira, Wadi, 34
hairy storks-bill, 101
Haj, Wadi, 31
Halal, Gebel, xii, 2, 30, 34
Halal Narrows, 31, 33, 34, 48
halophytic summer annual, 102
hammadas, 32
Hamman Faraun, Gebel, 20, 24, 43, 49
Hamr, Gebel, 44
hand of the merciful, 99
hand of the prophet, 99
hardpans, 83, 84, 86
Hart, Henry Chichester, 90
hawthorn, Sinai, 104
heat tolerance, 100
Hebran, Wadi, 37, 49
hidden water, 114
higher education, 8
higher plants, 92, 93
holy cities of Islam, 2
honey-dew mana, 94
hornets, 112
horticulture, 9
Host of Israel, 45, 112
hot springs, 49
house construction, 119
housing programs, 125
housing to put climate to work, 126
human ecology, 113
Hume, W.F., 19
Humr, Wadi, xii, 49
hyperarid areas, 60, 61
hyperaridity, 60, 61

Imlaha, Wadi, 49
Indian Mid-Ocean Ridge, 17
indurated layer (*see* hardpans)
Infra-Cambian Peneplain, 13
insects and arachnida
    beetles, 111

cicadas, 112
locusts, 112, 113
mosquitoes, 111
parasitic biters, 111
two-winged flies, 111
Insular Massifs, 2, 27–31
intertillage, 117
Iran Plate (Iranian Plate), 13, 16
irrigated agriculture, 81
Isla, Wadi, 49
Ismailia, x
Ismailia Fresh Water Canal, 118, 121, 123
Ismailia Governorate (*see* Al Ismailiyah)

jerboa, 105
Jethro, 45
Judeo-Christian shrines, 95
Jurassic sedimentaries, 20–22

Kassas, Mohammed, 88
Khatmia Pass, 29
Kheiyala, Gebel, 35, 40
Khizeimiya, Gebel, 44
Khodeir, Ahmed Ismail, vii
Kid, Wadi, 12, 49
Kid rocks, 12
Koppen/Rudloff climatic graphs, 65
Kotab, Hamdy, 87
Kuntilla, x, xii, 7, 50

Lakata, Wadi, 49
land bridge, 105
land mines, 75
large-leaf Siwa alfalfa, 117
Leja, Wadi, 43, 44
Letihi, Wadi, 37
Levantine Plate, 3, 11, 13, 16–19
Libni, Gebel, 3
Libyan Limestone (*see* Egma Limestone)
Lighthouse (*see* Madhawwa, Gebel)
limestone and dolomite, 22
limestone massifs, 29
limestone outliers, 26

literacy, 129
livestock diseases, 106
lizard tail (*see* shok-e-dhab)
low-energy period, 12
Lower Cretaceous uplift, 13
lower Paleozoic, 13
Lussa, Wadi, 34

Maaza Bedouin, 9
Madhawwa, Gebel, 41, 42
Maghara, Gebel, xi, 2, 3, 16, 22, 29,
    30, 48
Maghara Drainage, 47
magmatic intrusions, 17
Maharrad ruins, 42
malaria vector, 111
Mameluke rulers, 9
mammals
    Dorcas gazell, 106
    foxes, 113
    gazelle, 106
    Nubian ibex, 106, 108
    red fox, 113
    rock hyrax, 106
    Ruppels Fox, 113
mango grafting, 117
mangrove swamps, 101
manna, 94
Mansura, 121
Manzala, Lake, 121
marine deposits, 33
marine desert climates, 61, 62
Marine National Park, 131
Markha Plain, 35
massive folding, 12
maximum reproductive potential,
    115
Mecca, 8
Mediterranean
    Coast, 51
    Littoral, 2, 9
    Littoral Drainage, 47
    Sea, 3, 19
meeting place of humanity, 1
Mesozoic, 13
metamorphic materials, 20
metamorphics, 11, 12

methane, 129
microclimatology, 66
migmatites, 12
migration, human, 113
migratory locusts, 112
mining, 116
Miocene
    coral, 38
    rifting, 17
    sediments, 20, 21, 23
    subsidence, 17
    volcanic intrusions, 22, 23
Mir, Wadi, 44, 49
Mirad, Naqb, xi, 31
Mitla Pass, x, 16, 29, 31
Mitmetni, Wadi, 34
Mitmetni-Bedan Narrows, 33, 34
mizen (small clouds), 68
mobility, 113
Mohammed Ali, 10
Mokattam formation (*see* Eocene
    marls and shales)
Moon, F. W., 31
Moon God of Sumeria (*see* Sin)
Mother of Blackness (*see* Um
    Iswed, Gebel)
Mould of the Calf, 43
Mountain of Feran, 42
Mountain of Moses, 43
Mountain where God Spoke to
    Moses, 43
Mount Sinai, 43
Muarras, Gebel, 42
muhafazah (*see* governorates)
mult-variable climatic phenomena,
    59
Munsell color notations, 78, 85
Musa, Gebel, 39–41, 45, 46, 59, 95
Musa, Gebel (route to), 37

Na'ama Bay, 7, 8
Nabateans, 2, 114
Nakhaleh, Wadi, 42
Nakhl, x, 4, 7, 8, 32, 33, 67
narrow passes, 31
Nasb, Wadi, 49
Nasb-Dahab Wadi System, 38

nationalism, first, 114
Negev, 2
Nekhel (*see* Nakhl)
new agriculture, 116, 117
Nile Basin, population, 120
Nile sediments, 19, 24, 25
Noah's ark, 90
North Equatorial Trough of Low
    Pressure, 54
Northern Plains Drainage, 46, 48
North Sinai Governorate (*see* Sina
    ash-Shamaliyah)
Nubian sandstone, 22, 36
Nukhl (*see* Nakhl)
nummulitic limestones, 23
nurseries, plant, 118
Nusrani, Ras, 38
Nuweiba, 7, 8, 20, 35, 40, 50

ocean crust, formation, 18
Ogar, Gebel, 44
Oligocene sediments, 20, 21
Oligocene stratigraphic geology, 23
opportunism, economy of, 115
optimal water catchment, 98
organic matter (soil), 78

Palace of Abbas Pasha (*see* Qasr
    Abbas Pasha)
paleogeography, 13
Palmarian Folds, 18, 19
Palmer, Edward Henry, 4
Pangaea, 13
Paran, 2
Paran, Wadi, 50
parasitic symbiosis, 101
Pasha's Road, 45, 46
passes into the Dividing Valleys,
    37
pasture and forage, 118
Path of Moses, 46
Path of Our Lord Moses (*see* Sikket
    Syedna Musa)
pear grafting, 104
pedogenesis, 71
peninsula's greatest bane, 112
perennial grass, most important, 99

permanent housing, 125
permanent springs, 45
pH, 78, 79, 85
Phoenician juniper, 102
photovoltaics, 127
physical regionality, 2
physical regions, 2
Physical Sinai, ix
phytogeography
    African-wide groups, 90
    Asia-wide groups, 91
    Boreal Kingdom, 88
    chorotypes, 89
    climatic gradients, 96
    cosmopolitan plants, 90
    geologic factors, 89, 90
    geomorphological structure, 96
    Irano-Turanian Species, 89, 96,
        102–104
    limiting factors, 96
    macroclimatic factors, 95
    Mediterranean Species, 89, 98–
        104
    North African Region, 88
    North Egypt Region, 88
    Palaeotropical Kingdom, 88
    pantropical plants, 90
    protective mechanisms, 94
    regional desert species, 91
    regional interaction, 96
    regionally limited species, 91
    relict species, 91
    Saharo-Arabian Species, 89, 96,
        98–104
    Sinaitic regional plants, 91
    Sudanese Species, 89, 99–101,
        104
    Syria Region, 88
    upland and mountain plants, 94
    See also vegetative communities
Pilgrim Road (*see* Darb el Haj)
place names, xi, xii, 4
plant geography (*see* phyto-
    geography)
plate tectonics, 11
Pleistocene in Sinai, 24
Pliocene, 19

plunder, 31
plutons, 12, 20
Point 6 Ferry, 116
population
  growth, Nile Basin, 120
  increase, 114, 115, 120, 132
  relocation, 132
Port Said, x, 13
Port Said Governorate (*see* Bur
    Said)
Post, George E., 90
Precambrian
  granitics, 21, 36
  intrusions, 12, 20
  metamorphics, 12, 21, 36
  rocks of the Sinai Massif, 20
  uplift, 12
precipitation, 57–59
pressure, atmospheric, 53–55
Promulgation of the Law, 44, 95
    See also Rahah
pronatal sentiment, 115
protective cover, 105
pumping costs, 122
putting climate to work, 69

Qantara, x, 1, 5, 7, 8
Qa Plain, 3, 10, 27, 28, 35, 37, 38,
    49
Qasr Abbas Pasha, Gebel, 39, 40,
    44, 46
quartz-diorite, 12
Quartzipsamments, 71, 72, 75,
Quaternary
  alluvium, 21, 36
  central plateau, 25
  Gulf of Aqaba, 25
  Mediterranean Littoral, 25
  Qa Plain, 25
  shoreface plains, 25
  System, 24
Quseima, x, 7

Rafah, x, 1, 5, 7, 8, 29, 57, 58
Raha, Gebel, 23, 31
Rahah Plain, 43–45, 95, 122

rainfall variability, 59
rain-fed agriculture, 75
rains of Gemini, 68
Rakna, Naqb, 31, 35
Ras Mohammed, x, 1, 7, 8, 28, 37,
    39
Raz el Jeifi, 31
reclamation projects, 119–124
Red Sea, 3, 11
  Divergent Plate, 18
  Sea Mountains, 12
  Rift, 17
reefs, 37, 38, 101
refugial species, 102
reg, 31, 86, 96
regionally limited plant species, 91
relict forms, 102
reptiles and amphibians
  black mole viper, 105
  desert cobra, 109
  desert monitor, 110
  sand lizards, 110
  spiny-tail agama, 110
  whiptail lizards, 110
Research Center, Suez Canal Uni-
    versity, 44
research stations, 8, 129
resettlement policies, 114
retrograde deposition, Nile sedi-
    ment loss, 24
rifting, 17
Rimhan, Gebel, 40
rimth (*see* rope fiber)
ring-dikes, 12, 24
Ritter, Carl, 94
Romani, 7, 121
rope fiber, 99
rose of Jericho, 99
rotation of legumes, 118
Ruafu Dam, 34
Rudloff application of Koppen cli-
    matic system, 61–65
running water, imprint of, 26

Sa'al, Wadi, xii, 37
Sa'al rocks, 12

Sabkhet el Bardawil, x, xii, 24, 29
Safarial, Gebel (*see* bone bed)
sagebrush, 96, 101
Sahara, Gebel, 40
Sahbagh, Gebel, 39
Salaam Project, 87, 120–124
Salorthids, 72, 83
saltbush family, 118
salt deposition, 19
salt marsh habitats, 101, 102
salt tolerant crops, 117
Samr el Tinia (*see* Qasr Abbas
    Pasha)
sana'at (thunderheads), 68
San Diego, 67
sand valleys, 32–34
sayal acacia, 100
scuba, 8, 131
second-order soil survey, 70, 87
sedentarization, 119, 125, 126
sedges and tules, 102
sedimentary wedge, 20
seedstocks, 117
seheb (heavy rain clouds), 68
seil, xi
Senonian chalk, 20–22, 36
sensible heat, 52
Serabil, Gebel, 42
Serabit Khadim, x, 33, 35
Serbal, Gebel, xii, 12, 39–43
settlements, 5, 7, 8
seynite granite, 43
Shaesh, Moneer, 70, 87
shaggy sparrow-wort, 99
shallowing of the Mediterranean,
    19
Sharatib, Ras, 33, 35
Sharm el Sheikh, x, xi, 7, 8, 38, 51,
    58, 63
Sharm el Sheikh Research Station,
    8
Shefalla Fault, 36
Sheikh, Wadi, 44, 95
Sheikh Raiya, 37
Sheikh Zuwayid (*see* Rarah Plain)
Shibh Jazirat Sinai, 2

Shinenir, Gebel, 42
shittim wood, 90
shok-e-dhab, 104
shoreline erosion, 24
Shreich, Wadi, 46
Shulla, Gebel, 42
Siberian High Pressure System, 53
sidds (dry waterfalls), xi, 30
Sidri, Wadi, xii, 49
Sikket Shoeib, xii
Sikket Syedna Musa, 45
Silurian Period, 13
Sin, 2
Sina al-Janubiyah, 5, 6
Sina ash-Shamaliyah, 5, 6
Sinai
    area, 5
    gift of restless continents, 19
    incompletely severed umbilical,
        19
    Massif, 2, 4, 10, 12, 13, 26, 28,
        33, 39, 40, 50
    other names, 2
    population, 5
sink for upland drainage, 49
sinking (subsidence) Tenna Plain,
    24
small cattle, 4
small holdings, 123
small-plot agriculture, 117
Small Sea Canal, 121
smuggling, 8
snowfall, 59
soil
    classification (taxonomy), 71
    management, 81, 86
    map of World (UNESCO), 60
    mapping, 70
    regions of Sinai, 72
    at Son el Bricki, 73, 75–81
    surveys, need for, 70, 86
    temperature, 79
Solaf, Wadi, 35
solar cookers, 127
solar heating, 127
solar radiation, 51

Son el Bricki, 9
South Bardawil Basin, 122
South Sinai Governorate (*see* Sina al-Janubiyah)
speciation, levels of, 95, 96
spirit leveling (vertical triangulation), 46
Spring of Moses, 45
stable habitats, 29
St. Catherine, x, 7, 8
    Cluster (*see* Zebir, Gebel)
    Gebel, 12, 44, 46
    Monastery (Convent), 44, 45, 62
    Research Station, 8
stem-assimilant, 98
Stone of Moses, 43
Storie Index, 81, 82
subduction, 12
Sudr, Wadi, 49
Sudud, Wadi, 44
Suez, x
    Canal, 1
    Canal University, 8, 46, 129, 130
    Foreshore, 27, 28, 30
    Governorate (*see* As Suways)
Sufsafa, Ras, 43, 44
Sula, Wadi, 42
Sulaf, xii
summer cypress, 118
summer forage, 98
summer precipitation, 59
Summit Basin (Gebel Musa), 44, 45
summit of Serbal, 42
surface geology, 19–25
sustainable future, 113, 114
Suweir, Gebel, 40
syenite intrusion, 12
Syrian Christ-thorn, 100
Syrian sumac, 104

Taba, x, 7, 8, 13, 38, 50, 58
Taba-Nuweiba metamorphics, 20
Tarbush, Gebel, 39, 40
Tarfa, Wadi, 95
tarfa mana, 94, 95

Taurus Mountains, 18
taxonomy, soil, 79
temperature, atmospheric, 52
ten-feddan holdings, 123
Tenna Plain, 24, 120, 121
terracing, 81
Tethys
    Geosyncline, 13, 16
    Sea, 13, 16, 22, 105
Thamad, x, 7
Thamila, Naqb, 35
Thebt, Gebel, 39, 40
Thiman, Wadi, 49
third-order soil mapping, 70
thunderheads (*see* Sana'at)
Tih
    Escarpment, 31, 35, 48
    Gebel, 23
    Plateau, 2, 9, 26–28, 31–35, 57
toads, 111
tonalite, 12
toothbrush tree, 101
Tor, x, 5, 7, 62
Tor Drainage, 47
torric moisture regime, 73, 79
Torrifluvents, 80
Torriorthents, 72, 73
Torripsamments, 72, 73
tourism, 130, 131
tourist towns, 8
tourist village, 44, 131
Towara Bedouin, 10
triple-awned grass, 99
twisted-awn feather grass, 98

Um Adawi, 38
Um Adawi, Wadi, 49
Um Alawi, Gebel, 40
Um Bogma, x, 7, 13, 33, 35, 116
Um Iswed, Gebel, 43
Um Retama, Gebel, 24
Um Shomer, Gebel, xii, 12, 39–41, 46
Um Siyala, Gebel, 40
unchanneled flow, 33
UNESCO, 60, 71

universal thermal scale, 62, 63
Upper Eocene, 23
Uyun Musa, Gebel, 49

Van Plate, 16, 18
vegetative communities
  desert grasses with shrubs, 97,
    99, 100
  desert scrub, 96–98
  halophytic associations, 97, 101
  hot desert communities, 96–97
  interfluve communities, 96
  semishrubs, 97, 103
  semishrubs and grasses, 97, 103,
    104
  short-grass associations, 101
  short grass deserts, 97, 100, 101
  shrub and tree pseudo-steppes,
    97, 104
  shrub formations, 97, 102
  subdeserts, Mediterranean
    Coast, 97, 101
  subdeserts, mountain cores, 97,
    104
  subdesert uplands, 97, 102
  wadi communities, 97–99
vegetative formation, 95

wadi, xii
Wadi Akhdar Pass, 37
Wadi el Arish, 2, 32, 34
Wadi el Arish Drainage Basin, 31,
  48
Wadi Feiran metamorphics, 20
Wadi Feiran Oasis (*see* Feiran Oasis)
Wadi Kid metamorphics, 20
Wadi Nasb Pass, 40
Wadi Sa'al metamorphics, 20
wage labor, 115, 125
Wallace, Alfred, 105
Wardan, Wadi, 49
Wardan Drainage, 47
warmest station, 63
wasps, 112
waste conservation, 127–129
water budget climatology, 63, 66, 67

Watering Place, Pass of (*see* Mirad,
  Naqb)
waterlogging, 122
Water Resource Agency of Sinai, 87
Watia Pass, x, 35, 40
Watir, Wadi, 26, 35, 37, 50
Watir Drainage, 47, 49
Way to Hejaz, 2
Western Marginal Faults, 36
white-branched orache, 98
white broom, 98
wide ranging plants, 90
Wilderness of the Wandering, 31
Wilderness of Tih, 40
winds
  Hamasin, 56, 57
  prevailing, 56
  Shamal, 57
windsurfing, 8, 132
wine, 8
winter, 68
winter rainfall, 59
World Bank, 87
World Food Program
  arresting environmental degrada-
    tion, 117
  fodder shrubs, range develop-
    ment, 117, 118
  fruit tree plantings, 117
  land reclamation, 117, 119–124
  training centers, 118
wormwood, 96, 101

xeric moisture regime, 60, 61, 79
Xerofluvents, 72, 73, 79
xerohalophyte, 96
xerotropic forms (animals), 106

Yamit, 7
Yelleq, Gebel, 2, 29, 48
Yelleq Massif, 33
yellow fever vector, 111

Zaghra, Wadi, 23
Zebir, Gebel, 39–41, 44, 46, 59
Zelega, Wadi, 33, 35, 48

Zelega-Biyar-Ain wadi system, 37
Zohary, M., 90
zoogeography, 104–113
    biological isolation, 106
    burrowing, 105
    crustal tectonics, 105

Desert Belt, 105
global desiccation, 105
Mediterranean Subregion, 105
Palearctic, 105, 106
protective cover, 105